PIONEERS IN POLYMER SCIENCE

CHEMISTS AND CHEMISTRY

A series of books devoted to the examination of the history and development of chemistry from its early emergence as a separate discipline to the present day. The series will describe the personalities, processes, theoretical and technical advances which have shaped our current understanding of chemical science.

PIONEERS IN POLYMER SCIENCE

by

RAYMOND B. SEYMOUR
HERMAN F. MARK
LINUS PAULING
CHARLES H. FISHER
G. ALLAN STAHL
L. H. SPERLING
C. S. MARVEL
CHARLES E. CARRAHER, Jr.

Edited by

RAYMOND B. SEYMOUR

Department of Polymer Science,
The University of Southern Mississippi, U.S.A.

KLUWER ACADEMIC PUBLISHERS
DORDRECHT / BOSTON / LONDON

Library of Congress Cataloging in Publication Data

```
Pioneers in polymer science / by Raymond B. Seymour ... [et al.] ;
  edited by Raymond B. Seymour.
      p.   cm. -- (Chemists and chemistry)
  Includes bibliographies and index.
  ISBN 0-7923-0300-8
  1. Chemists--Biography.  2. Polymers--History.  3. Plastics-
  -History.   I. Seymour, Raymond Benedict, 1912-   . II. Series.
TP1119.A1P56   1989
668.9'092'2--dc20                                          89-8222
```

ISBN 0-7923-0300-8

Published by Kluwer Academic Publishers,
P.O. Box 17, 3300 AA Dordrecht, The Netherlands.

Kluwer Academic Publishers incorporates
the publishing programmes of
D. Reidel, Martinus Nijhoff, Dr W. Junk and MTP Press.

Sold and distributed in the U.S.A. and Canada
by Kluwer Academic Publishers,
101 Philip Drive, Norwell, MA 02061, U.S.A.

In all other countries, sold and distributed
by Kluwer Academic Publishers Group,
P.O. Box 322, 3300 AH Dordrecht, The Netherlands.

printed on acid free paper

PREFACE

Because of a lack of appreciation for his efforts in developing modern polymer science, the contributions of Hermann Staudinger were disregarded for decades. There have also been delays in recognizing the contributions of other pioneers in polymer science.

Hence, it is gratifying to note that Professor Seymour chaired an American Chemical Society Symposium focusing on the contributions of these pioneers and that Kluwer Academic Publishers has published the proceedings of this important symposium.

H. Mark

DEDICATION

This book on Pioneers in Polymer Science is dedicated to Nobel Laureate Polymer Scientists Hermann Staudinger, Emil Fischer, Herman Mark, Paul J. Flory, Linus Pauling, Carl S. Marvel, M. Polanyi, Giulio Natta, Karl Ziegler, and Bruce Merrifield as well as to those pioneers such as J.C. Patrick, Robert Thomas, William Sparks, Maurice Huggins, Otto Bayer, Leo Baekeland, Anselm Payer, Roger Boyer, Waldo Semon, Robert Banks, J.P. Hogan, and other pioneers who, to a large degree, were responsible for the development of the world's second largest industry.

ACKNOWLEDGEMENT

The editor appreciates the contribution of co-authors Herman Mark, C.H. Fisher, and G. Alan Stahl who co-chaired the Symposium on Pioneers in Polymer Science at the National Meeting of the American Chemical Society at Seattle, WA in 1984 and who contributed a chapter in this book.

The editor is particularly grateful to Mischa Thomas who typed this manuscript.

Raymond B. Seymour
Distinguished Professor

TABLE OF CONTENTS

CHAPTER 1

PRE-TWENTIETH CENTURY POLYMER PIONEERS

ABSTRACT

The ACS Center for History of Chemistry (CHOC) emphasized the history of polymer chemistry (1920-1961) in its first major project. Some polymer history has been discussed at ACS National Meetings at Houston, Las Vegas, Seattle, and Washington, but the emphasis has been on 20th century contributions. Emphasis in this chapter will be on early contributions to polymer science by Simon, Regnault, Berzelius, Goodyear, Braconnot, Pelouze, Gough, Schonbein, Menard, Kekule, Lourenco, Parkes, Hyatt, Piutti, Hlasiwetz, Chardonnet, Musculus, Critchlow, Brewley, Brown, Lintner, Von Pechmann, Baumann, and Schulzenberger. These early investigators contributed both to the science and technology of plastics, fibers, and elastomers and set the stage for more significant developments in the 20th century.

Most history of chemistry books include descriptions of the arts of making wine, vinegar, pottery and glass as well as metallurgy and dyeing . The art of food preparation, caulking of ships, spinning and weaving of textiles (1), tanning of hides, painting (2) and waterproofing with natural rubber (3) are infrequently mentioned by many historians. Yet, like the dyeing of wool and other textiles, these arts were at least as important for the food, shelter, clothing and enjoyment of the ancients as were the more highly publicized arts of pottery making and metallurgy.

Yet, the importance of vegetable and animals and some of the endeavors of these animated polypeptides are overlooked by non-scientists and scientific historians alike. In 1816, Christian Thomsen, a Danish museum curator,

1

R. B. Seymour (ed.), Pioneers in Polymer Science, 1–11.
© *1989 by Kluwer Academic Publishers.*

coined the terms stone, bronze, and iron ages for the three prehistorical periods. The general use of these terms has helped to subordinate the importance of proteins, polysaccharides, nucleic acids and polymeric hydrocarbons which existed before the stone age and are still in existence today.

Caulking of the ark with pitch is described in Genesis 6.14. Linen fabric, at least 7000 years old has been unearthed near Robenshausen, Switzerland. Wool fabric was used in Mesopotamia as early as 4000 B.C., cotton was woven by the ancient Egyptians and Sericulture was practiced in China as early as 2500 B.C.

Leather made by the crosslinking of proteinaceous hides (tanning) has been called "the most historic of useful materials". Reference to the tanner was made by the playwright Aristophanes more than 2,500 years ago.

The crude art of painting, using earth colors and vegetable dyes, as well as proteinaceous (egg white and milk) or resinous binders (pitch), was practiced over 4,000 years ago.

Ancient artisans also chiseled, cut and formed tortoise shell in addition to chiseling stone, ivory, animal hooves and horns. The word plastics was used in a mold patent in 1862 and in spite of an attempt by the National Electric Manufacturer's Association (NEMA) to change it to "synthoid", the word plastic continues to be used. The original chiselers were called horners but the name plastics technologists or engineers has now displaced this restrictive term in the USA.

Cachuchu ("the wood that weeps") was used for recreating, religious rites and waterproofing by the American Indians long before Columbus landed in the

West Indies in 1492. As is the case with other ancient polymeric art, few names could be associated with this elastomer prior to the eighteenth century. Joseph Priestly discovered that the cachuchu could be used as an eraser for pencil marks and coined the name rubber for *Hevea braziliensis* in 1770. The word rubber continues to be used in many english-language countries.

Thomas Hancock built a factory for producing rubber articles in 1820 and Charles MacIntosh obtained waterproof fabric by making a rubber-cloth sandwich in 1823. It is of interest to note that John Jacob Berzelius coined the word polymer in 1833 just seven years after Michael Faraday had shown rubber to have the composition of C_5H_8 and six years before the discovery of the vulcanization process by Charles Goodyear. Berzelius also coined the terms isomer, catalyst and protein (4) and synthesized glycerly tartrate polyester resins (5).

A contemporary, Henri Braconnot, produced Xyloidine in 1833 by the reaction of nitric acid and starch and another contemporary, Theophile Jules Pelouze produced "Pyroxyline" by the reaction of nitric acid and paper in 1838. Subsequently, in 1836, Christian Fredrich Schonbein obtained a higher degree of substitution in cellulose by using a mixture of nitric and sulfuric acid. Schonbein obtained patents in England and USA for his "Nitrocellulose" which he called "gun cotton".

Other contemporaries, Henry Victor Regnault and E. Simon, described polymers of vinyl chloride, vinylidene chloride, styrene and other polymers in 1938 and 1939. However, the most significant contribution to polymer science in the early part of the nineteenth century was made by a blind scientist named John Gough.

In a letter to Dr. Holmes in 1802, John Gough described results of three experiments which formed the basis for the modern theory of rubber elasticity (5):

1.   that a piece of rubber became warm to the lips when stretched and this thermal effect was reversible.

2.   that a stretched piece of rubber contracted when heated.

3.   that rubber retained it's reversible elasticity in warm water but lost some of its retractile properties in cold water.   However, extended rubber contracts rapidly when heated.

James Prescott Joule measured the increase in temperature when rubber was stretched in 1859.   William Thompson (Lord Kelvin) was one of the few natural philosophers (scientists) to recognize the importance of Joule's investigations.   He developed an equation for quantifying Joule's discovery in 1857.   Units for temperature and energy have been named after these polymer pioneers.

Natural rubber was crosslinked or vulcanized by Liedershoff who heated it with sulfur in 1832.   In 1838, Nathaniel Hayward patented the use of sulfur as an additive to decrease the tackiness of natural rubber.   After purchasing this patent, Charles Goodyear accidently allowed this patented mixture to remain overnight on a hot stove and obtained a more versatile product which he called "Vulcanite".

Goodyear did not apply for a patent for vulcanization "low density crosslinking" of rubber until 1843.   Prior to 1843, but after Goodyear's discovery in 1839, Thomas

Hancock patented the sulfur curing process for rubber in England.

Hancock made solid rubber tires for Queen Victoria's carriage in 1846 and was moderately successful in his English rubber business. However, Robert William Thompson had previously invented pneumatic rubber tires in 1845 and these were reinvented by John Boyd Dunlop in 1888.

Charles Goodyear wasted much time and money defending his vulcanization patent which issued in 1844 while he was in debtor's prison. His attorney Daniel Webster, was successful in "The Great India Rubber patent infringement case" which established the validity of Charles Goodyear's patent in 1852. Goodyear displayed his knowledge of rubber elasticity by authoring two books on vulcanized rubber. His brother Nelson Goodyear displayed at least a superficial knowledge of degrees of crosslink density when he produced hard rubber (ebonite) by heating rubber with 40-50 parts per hundred of sulfur. Nelson's patented thermoset polymer was the first manmade plastic. Thomas Hancock also patented hard rubber in 1843. Ebonite was used to make the "fountain pen" which was patented by L.E. Waterman in 1884.

Celluloid, the first manmade thermoplastic, was not invented until after the 1850's. J. Parker Maynard obtained a solution which he called Collodion by dissolving cellulose nitrate in an equimolar mixture of ether and ethanol in 1847. In 1862, Alexander Parkes patented the process of coating fabric with Collodion. He also shaped this cellulose derivative, which he called Parkesine, under heat and pressure, and was awarded a bronze medal for this invention at the Great International Exhibition in London in 1862.

F. Scott Archer used Collodion for photographic film and Schonbein used this solution to provide waterproof paper. One of the most important contributions was the addition of camphor by J. Cutting who patented this composition in 1854.

After the failure of the Parkesine Company, Parkes and Daniel Spill formed the Xylonite Company which produced cellulose nitrate plastics under the trade names of Xylonite and Ivorude. Camphor, which Parkes claims to have used as a plasticizer or solvent was mentioned in his patent in 1855. Daniel Spill patented compositions containing cellulose nitrate and camphor in 1869.

Several English and American inventors obtained patents on applications of Pyroxiline plastics. The most significant patents were those obtained by John Wesley Hyatt for molding mixtures of Pyroxiline and camphor under heat and pressure in 1869 and in the 1870's (7). His brother, Isiah Smith Hyatt coined the name "Celluloid" and registered this trademark in the U.S. Patent Office.

While Celluloid was developed originally for use in place of ivory for making billiard balls and used later for dentures and shirt fronts, the annual volume never exceeded 500 tons during the nineteenth century. Much of this volume was used for photographic film, hairpins and combs.

The Hyatt brothers did little to advance polymer science but they did influence the development of Charles Burrough's compression sheet molder and blow molding press. They also utilized H. Brewley's extruder. The latter was developed originally in 1845 for the extrusion of shellac and gutta percha (8).

Alfred P. Critchlow, another plastic molding pioneer, molded shellac and gutta percha in Florence, MA. His firm is now known as the Prophylactic Company. Many of these developments occurred in the area around Leominster which became a center for plastic machinery and plastic fabrication (9). Leominster is also the sight of a national plastic museum.

Hannibal Goodwin patented the use of Celluloid photographic film in 1887, John Henry Stevens and Marshall C. Lefferts patented a machine which made films by coating solutions of cellulose nitrate in amyl acetate (lacquer) in 1892.

In 1664, Robert Hooke predicted that fibers equal to those produced by silkworms would be made by mechanical means. Joseph Wilson Swan produced filaments from Collodion in 1883 and attempted to use these filaments to produce carbon filaments for incandescent lamps. His earlier investigations of carbon filaments for illumination in 1860 preceded similar developments by Thomas Edison in the 1880's. Swan was knighted for his invention in 1904.

Count Louis Marie Hilaire Chardonnet, an assistant to Louis Pasteur, patented the process of producing filaments by forcing Collodion through small holes (spinnerets) in 1884. This "Chardonnet silk" was a sensation at the Paris Exposition in 1891. Because of its inherent flammability, this fiber was called "mother-in-law silk". Nevertheless, Chardonnet received the Perkin Medal in 1914 for this development. The carbon fibers used by Swan & Edison in the nineteenth century were also used a century later as reinforcements for sophisticated plastic composites.

L.H. Despaissis patented the cuprammonium process for the production of regenerated cellulose (rayon) from a solution of cellulose and cuprammonium hydroxide (Schweitzer's reagent) in 1890 (10). Filaments were produced from these solutions by Fremery and Urban in Germany at about the same time (11). Schweitzer's reagent continues to be used in laboratories for viscosity tests but the use of this solution for rayon production has been discontinued.

Charles Frederich Cross produced regenerated cellulose by the acidification of an alkaline cellulose xanthate solution in 1892. This process was patented by C.F. Cross, E.J. Bevan and C. Beadle in England in 1893. However, the commercial viscose process, which is still in use today, was not developed until 1903. The name rayon was accepted by all producers of regenerated cellulose fibers in 1924.

M.P. Schutzenberger produced cellulose acetate, with a degree of substitutions (DS) of about three by the reaction of acetic anhydride and cellulose in 1859. Cross and Bevan produced filaments and films from chloroform solutions of this triacetate but this was not an economical process. The commercial cellulose acetate rayon fiber process, based on acetone solutions of secondary cellulose acetate (DS=2), was developed and patented by G.W. Miles who partially saponified the triacetate in 1903 (11).

Natural rubber was chlorinated by G.A. Englehard and H.H. Day in 1859, hydrochlorinated in 1881 and isomerized by Leonhardi in 1881 (12). However, these rubber derivatives were not commercialized until the early part of the twentieth century. Likewise, polyvinyl chloride was described by Regnault in 1838, by Meyer in 1866 and by Baumann in 1872 (13,14) but was not commercialized until the 1930's. It is of interest to note that Baumann considered the solid polymer to be an isomer of vinyl chloride.

The polymerization of styrene and acrylic acid was observed in 1872 but the polymers of these monomers were not produced commercially until the early part of the twentieth century. Likewise, Gay-Lussac and Pelouze in 1833, Berzelius in 1847, Bemmelen in 1856, Lourenco in 1863 and Watson Smith in 1899 investigated the formation of polyesters but these polymers were not commercialized until the early part of the twentieth century.

Advances in the nineteenth century were limited by economics and lack of knowledge of polymer science. Kekule did suggest that proteins, starch and cellulose consisted of very long chains and Lourenco used the term "highly condensed" to describe his polyesters (15) in 1863. Likewise, Hlasiwetz and Habermann used the term soluble and unorganized to differentiate small molecules from insoluble organized natural polymers in 1871.

Other investigators in the nineteenth century including Emil Fischer, reported unusually high molecular weights for natural polymers. Nevertheless, confusion between concepts of colloids, the association theory and that of macromolecules continued to retard progress until Nobel Laureate Hermann Staudinger championed the true macromolecular concept in the early 1920's.

As pointed out in this introductory chapter, some progress was made and the foundations for modern plastics, fiber, coatings and rubber industries were laid prior to the twentieth century. However, much of this progress was retarded and industrial growth was hampered because of the need to rely on an empirical rather than a scientific approach. While this phase of polymer development has been appropriately called "ignorance in action," it should be noted that all progress in polymer science and technology was made by a small number of artisans and entrepreneurs and even a smaller number of scientists.

The beginning of the "golden age" of polymers was delayed until scientist accepted the concept of macromolecules, as described by Staudinger in the 1920's and 1930's. As a result, Staudinger and each of his followers accomplished more than was accomplished by the combined efforts of all the investigators prior to the twentieth century.

REFERENCES

1. C.H. Fisher, Chapt. 16 in "History of Polymer Science and Technology," R.B. Seymour ed., Marcel Dekker Inc., NY, 1982.

2. R.R. Myers, Chapt. 5 in "History of Polymer Science and Technology," R.B. Seymour ed., Marcel Dekker Inc., NY, 1982.

3. P.E. Hurley, Chapt. 13 in "History of Polymer Science and Technology," R.B. Seymour ed., Marcel Dekker Inc., NY, 1982.

4. G.A. Stahl, Chapt. 3 in "Polymer Science Overview," G.A. Stahl ed., American Chemical Society, Washington, D.C., 1981.

5. R.B. Seymour, Chapt. 7 in "History of Polymer Science and Technology," R.B. Seymour ed., Marcel Dekker Inc., NY, 1982.

6. G. Allen, Chapt. 1 in "Pioneers of Polymers," The Plastics and Rubber Institute, London, 1981.

7.   R. Friedel, "Pioneer Plastics," "(Celluloid)", The University of Wisconsin Press, Madison, WI, 1983.

8.   J.H. DuBois, "Plastics History USA," Cahner Books, Boston, MA, 1972.

9.   W.A. Emmerson, "Leominster, Historical and Picturesque," Lithotype Publishing Company, Gardner, MA, 1888.

10. H.L. Herger, G.C. Daul, Chapt. 1 in "Solvent Spun Rayon, Modified Cellulose Fibers and Derivatives," A.F. Turbak ed., American Chemical Society, Washington, D.C., 1977.

11. P.W. Morgan, Chapt. 4 in "History of Science and Technology," R.B. Seymour ed., Marcel Dekker, Inc., NY, 1982.

12. R. Lenz, Chapt. 17 in "Organic Chemistry of Synthetic High Polymers," Interscience Publishers, NY, 1967.

13. W.L. Semon, G.A. Stahl, Chapt. 12 in "History of Polymer Science and Technology," R.B. Seymour ed., Marcel Dekker Inc., NY, 1982.

14. M. Kaufman, "The History of PVC," Gordon and Breach, NY, 1969.

15. G.A. Stahl, Chapt. 3 in "Polymer Science Overview," American Chemical Society, Washington, D.C., 1981.

CHAPTER 2

DEVELOPMENT OF A MODERN POLYMER THEORY

G. Allan Stahl
EXXON Research & Engineering Co.
Research and Development
Baytown, TX 77501

ABSTRACT

When presenting the 1953 Nobel Prize in Chemistry to H. Staudinger, A. Fredga said, "In the world of high polymers, almost everything was new and untested. Long standing concepts had to be revised or new ones created. The development of macromolecular science does not present a picture of peaceful idylls." The model of a high polymer consisting of long covalent chains had its roots in nineteenth century structural organic chemistry. Acceptance of the model was, however, sidetracked in the early 1900's by the development and popularization of colloid science. In 1920, the characteristic properties of high molecular weight substances was explained by "association" forces. In that year Staudinger published his classic, "Uber Polymerization". Publication of this paper heralded a decade of intense research and, at times, even more pitched controversy. The story of the development of the modern polymer theory is an excellent case study in the formation and confirmation of a hypothesis. The story is made spicy by the intensity of the debate. One participant recently commented that the champions of the long chains often argued with each other more vigorously than with defenders of the association theory.

R. B. Seymour (ed.), Pioneers in Polymer Science, 13–28.
© 1989 by Kluwer Academic Publishers.

INTRODUCTION

At least fifty percent of the chemists in the United States are employed in the preparation, characterization, or application of polymeric materials. Billions of dollars of commerce and reams of pages of research reports which are produced annually are spurred by the diverse physical properties of polymers.

These physical properties can be classified in a myriad of ways including solvency, hardness and reactivity. The chemical constituents of the backbone and pendant groups affect these properties. But it is the common chemical property, high molecular weight through repeating covalent bonds, which sets polymers apart from all other kinds of materials. High viscosity, elasticity, and strength are a few of the consequences of high molecular weight.

The existence of large molecules with linear molecular weights of thousands, even millions, is now accepted by almost everyone. Even elementary-aged children learn of DNA and the size of its strands. High molecular weight is an accepted building block today yet, this was not always true. The existence of high molecular weight materials was accepted in all scientific quarters only in the 1930's.

Although Berzelius had not considered high molecular weight, his definition of a polymer contained essential elements. His concept, unchallenged in the mid-nineteenth century, was strengthened by August Kekules establishment of the importance of structure to chemistry. In 1878, Kekule went so far as to propose that polyvalent atoms might form "sponge or net-like" molecular masses.

Workers in this era frequently used the word polymer, but not always in the sense that it is used today. Two groups, knowingly or not, actually studied high molecular weight substances. They were (1) those concerned with the chemical and physical properties of natural products and (2) the synthetic organic chemists (especially in Germany) who served as the vanguard of the modernization of organic chemistry. Hindsight makes one important point clear. The path to the acceptance of the modern polymer theory would have been less torturous had the former recognized the significance of the occasionally reported synthesis of high molecular weight substances by the latter.

An early example of a deliberate polymer preparation can be found in the work of A.V. Lourenco (1). In 1860, he reported the preparation of a series of adducts of ethylene glycol and tolylene dichloride. Isolation and identification showed several members of a homologous series with degrees of polymerization (DP) of 2 to 6. He noted the boiling points increased with DP. It was with great insight that Lourenco predicted that highly viscous, undistillable products, obtained with more drastic conditions, were of "a higher degree of condensation". Lourenco reported that ethylene glycol-succinic acid adducts were also highly condensed.

The idea that proteins and carbohydrates are polymeric goes back to the work of H. Hlasiwetz and J. Haberman (2). In 1871, they proposed that these substances were made up of a number of species differing in their degree of condensation. More important, they differentiated between "the soluble and unorganized" members of the groups, such as dextrin and albumin, and the "insoluble and organized" members, such as cellulose and keratin. This distinction was the precursor of the present day differentiation, of non-crystalline and crystalline polymers.

Workers in many quarters reported high molecular weight substances; their estimates of molecular weight were often in the ranges of 30-50,000. As if to give final confirmation to the concept, the eminent Emil Fischer turned to the study of polypeptides in his first years at Berlin. With characteristic thoroughness, he described numerous polypeptides, eventually preparing one with a molecular weight of 4200. Fischer, in fact, never postulated any structure for these materials other than an amino acid with a repeating CO-NH link.

In 1906, he proposed an uninterrupted series between the simple di- and trimers of amino acids and proteins. In 1913, he reemphasized his beliefs by stating that the "ultimate proof of high molecular weight came from his synthetic products made by analogous, controlled chemical reactions".

A NEW CONCEPT-COLLOIDAL SUBSTANCES

Near the middle of the nineteenth century, Thomas Graham noted that certain materials, including many gums, did not diffuse through certain gelatinous substance. He dubbed these materials "colloids". Graham's definition was expanded until the concept developed that a substance could exist in a colloidal state just as it could in gaseous, liquid, or solid state. This expanded definition wouldn't enter our narrative had its developers accepted the possibility of high molecular weight. Many scientists, however, chose to ignore this possibility and a scientific misadventure occurred.

Those studying colloids expanded their concepts until in 1899 Johannes Thiele extended his valence theory to double bonds. He suggested that the material, now known as polystyrene, was merely styrene bound by association of its double bonds. he called this "partial valence". This view of colloidal properties through partial valences (and not through high molecular weight) was known as the association theory.

Many noted scientists supported the association theory. Among them (and important to the story of the development of the modern polymer theory) were H. Pringsheim and K. Hess who applied the concepts to explain the properties of cellulose just as M. Bergmann and E. Abderhalden had done with proteins (3).

When challenged, supporters of the theory quickly rejected any possibility of materials with high molecular weight. They cited supportable experimental results, and at times evoked elaborate rationalizations. It is only fair to point out that with 70 years of hindsight, H. Crompton's following argument is amusing but it was tendered with thought and sincerity;

"No upper limit is usually assigned to molecular weight. E. Fischer has synthesized a polypeptide with the molecular weight of 1212, and in the case of colloids, molecular weights in the order 10^4, and even 10^5, are commonly spoken of. A difficulty arises, however, in admitting that molecular weights can exceed a certain value, unless the density increases as the molecular weight increases.

For suppose, that a compound such as a protein can exits, with a density not much greater than that of water, and with a molecular weight of more than 30,000, the grammolecule of such a compound at 0°C would occupy about 30,000 cc. The grammolecule of a perfect gas under the standard conditions occupies only 22,400 cc, and we should therefore have a solid compound, at 0°C and under a pressure that cannot be less than one atmosphere, occupying a greater molecular volume than that of any gas..."

Two suggestions appear to be indicated. The first is that under the ordinary conditions, there is an upper limit to molecular magnitude, and that for most substances, especially colloids, the molecular weight cannot exceed a value of about 20,000. The second is that our ordinary kinetic-molecular conceptions no longer apply when for a given temperature, the molecular magnitude exceeds a certain critical value. The latter view is in accord with our present knowledge and perhaps serves to throw some light on the behavior of colloids (4)."

The acceptance of the association theory and rejection of the possibility of high molecular weight was supported by three developments:

1. Colloid science was found useful in bridging the gap between biological and physical science. Thus, it received quick and reputable recognition.

2. The existence of partial valences was explained in Alfred Werner's proposed "hauptvalenzen" (primary valence) and "Nebenvalenzen" (secondary valence).

3. X-ray crystallographers were unable to explain the diffraction patterns of high molecular weight substances. They accepted, as fact, that a molecule could be no larger than its unit cell.

As the association theory became ensconced in the minds of the scientific community, the high molecular weight concept regained the path to final acceptance. It bounced back with studies of the structure of rubber.

RUBBER

The history of the use and study (it occurred, as in so many other cases, in that order) of rubber goes way back. Pietro Martyre d'Anghiera described it in 1511. In 1839, Charles Goodyear found that sulfur could vulcanize rubber, thus its usefulness was greatly expanded. Still rubber, except for Faraday's report on its empirical composition, was an unknown, until Greville Williams destructively distilled the material in 1860 and identified the distillate as isoprene.

Synthetic chemists joined in at this point. Gustave Bouchardat (1879) (5) prepared a rubber-like material by the action of Cl_2 or HCl on isoprene which "dissolved after the fashion of natural rubber"; William Tilden (1882) (6), Otto Wallach (1887) (7), and J.H. Gladstone and William Hibbert (1889) (8) also made elastomers from isoprene. The latter workers reported a controversial molecular weight of 12,000 in 1889.

It was Harries who (by ozonolysis) demonstrated that rubber contained a repeating isoprenic structure which was capable of chain expansion by addition of isoprene units. Since he was unable to detect end groups, Harries proposed ever expanding "rings" of isoprene. Conforming somewhat to the wisdom of the day, he expounded the idea that the rubber-like properties were caused by aggregation of the rings.

In 1910, S.S. Pickles (9) proposed that rubber was made of long chains, and that the elastic properties were based on chain lengthening. Shortly thereafter, Harries adopted this concept and began talking about ever lengthening chains of 5, then 7, then 9 isoprene units.

STAUDINGER AND A "NEW" THEORY

Herman Staudinger also studied the polymerization of isoprene in 1910. His studies sparked an insight into the makeup of all materials that we now recognize as polymers. An established and highly reputable organic chemist, Staudinger, caused quite a stir in the old German school when he redirected his research to study these substances.

Most polymer preparations, in the nineteenth century, were isolated events. For example, vinyl chloride and styrene were gelled by the action of sunlight. Undaunted, Staudinger began the systematic preparation of various adducts of polymerizable monomers. Careful and thoughtful study led to the conclusion that the association theory was wrong.

In 1920, he published his classic paper "Uber Polymerization" (10) in which he summarized his findings and proposed formulas for polystyrene and polyoxymethylene which are accurate even today. He even suggested a chain structure for rubber and claimed that its properties were due entirely to high molecular weight.

About the same time that Staudinger was publishing his findings and proposing high molecular weights, R.O. Herzog and W. Jancke at the Kaiser Wilhelm Institute showed that at least part of cellulose was crystalline. Employing x-ray diffraction, they obtained powder patterns which were neither clear spots nor powder rings. Unable to decipher the diagram, Herzog assigned the task to Michael Polyanyi. Polanyi's interpretation followed Staudinger's paper as the second major step to the resolution of modern polymer theory. Polanyi concluded that the diffraction diagram was in agreement with either long glucosidic chains or dimer rings. Immediate response by the scientific community was to ignore the first possibility. Developments in the 1920's proceeded from this point along two lines; Staudinger's chemical investigations and x-ray diffraction studies.

Between 1922 and 1930, Staudinger published 19 papers on rubber alone and continued the systematic preparation of many other polymers. He hydrogenated rubber, but, the failure of this hydrogenation to alter the "colloidal properties" was not accepted, as proof of high molecular weight!

The division of ideas between Herzog and his assistants, Polyanyi, Herman Mark, Karl Weissenberg, and Rudolf Brill grew in this period. In 1923, Brill produced x-ray diagrams of silk fibroin and concluded that there were eight amino acid residues in the unit cell. In 1925 Weissenberg openly discussed the possibility of high molecular weight. It is to Herzog's great credit that, although he accepted only a degree of polymerization of "two or a slightly higher number", the work went on.

About 1925, J.R. Katz (11) noted, in a now famous experiment, that stretched rubber contained partially crystallized material, which was supportive to the concept of high molecular weight. E.A. Hauser and Mark stressed this concept in their quantification of Katz's work in 1926. The same year, O.L. Sponsler and W.H. Dore presented a complete description of the cellulose molecule. Making use of Haworth's deduction of 1,4 bonded glucose rings, they proposed a chain of glucose rings joined by covalent bonds.

A STORMY DECADE

When Herman Staudinger received the Nobel Prize in 1953, for his work on macromolecules (Staudinger dubbed the term), A. Fredga said:

Professor staudinger.   Thirty years ago, you adopted the view that a Chemical molecule is able to reach almost any size... It is no secret that for a long time many colleagues rejected your views which some of them even regarded as abderitic. Perhaps this was understandable. In the world of high polymers, almost everything was new and untested.   Long standing established concepts had to be revised or new ones created. The development of macromolecular science does not present a picture of peaceful idylls (12).

Fredga was referring, in part, to the debate between the advocates of high molecular weight and the association theorists. But an equally intense debate raged within the camp of the new concept. These conflicts came to a head in the years 1925 to 1930. A glimpse of the stormy events can be seen on the occasion of Staudinger's farewell address to the Zurich Chemical Society. He lectured on the existence of long thread-like molecules consisting of Kekule or covalent bonds. His ideas were in direct conflict with some x-ray crystallographers' concept of a small unit cell corresponding to a low molecular weight. Many noted scientists, including Karrer, Niggli, Wiegner, Ott and Scherrer, tried to convince him that his concepts conflicted with exact scientific data. It has been reported that the meeting ended abruptly when Staudinger pounded his fist and said, "Here I stand, I cannot move!"

Aware of growing interest in the debate, Richard Willstaetter arranged a symposium at the "Gesellschaft Deutscher Naturforscher und Arzte: in 1926 (13). The meeting brought Staudinger face-to-face with his antagonists; Bergmann, Pringsheim, and E. Waldschmidt-Leitz. The fifth speaker was an x-ray crystallographer, Herman Mark, who at the urging of Fritz Haber, was invited by Willstaetter. Mark was asked to address specifically the issue of molecular weight and the unit cell.

Bergmann and Pringsheim opened with arguments on "high molecular weight" inorganic complexes and inulin (a polysaccharide now known to have a molecular weight of about 5000). These examples and the citation of numerous scientific leaders led them to conclude that the properties of many compounds were due to the combination of primary and secondary valences and that the properties (now attributed to high molecular weight) were a result of the latter. Staudinger presented voluminous data on polymerization, hydrogenation, viscosity, melting points, and solubility. He concluded by reporting on the conversion of polystyrene to polyhexahydrostyrene and polyindene to polyhexahydroindene with no loss of properties. He maintained that this proved these monomers were united by covalent bonds.

His point might have been lost, but for the presentation of data by x-ray crystallographer Mark. Addressing the issue of the unit cell, Mark concluded, "Cellulose is united by forces comparable by type and magnitude to the inner forces of the molecule and the whole crystallite appears as a large molecule."

Shortly after this meeting in 1927, Mark wrote Staudinger and asked for a sample of polyoxymethylene for his x-ray investigations. Staudinger declined since he was already working with J. Hengstenberg on this compound. The two, in fact, corresponded through 1927.

In 1928, K.H. Meyer and Mark published an important, extensive paper on cellulose (14). They proposed a structure which is comparable to that accepted today. They proposed a model of amorphous cellulose bark with high molecular weight cellulose chains linking the bark with more ordered, highly crystalline regions. Employing both primary and secondary bonding, the Meyer-Mark model was a compromise between the association theory and Staudinger's macromolecule. Staudinger disagreed with Meyer and Mark on two points. He felt that their main chains were not long enough and disagreed with the concept of linked amorphous and crystalline regions.

Their disagreement was slight when Meyer reiterated their views in a second paper. Shortly afterward, in October, 1928, Staudinger formally criticized their views and dubbed their model "the new Micelle Theory" (15). The most important issues to Staudinger were the importance of secondary bonding to physical properties and priority.

It is important to note that, in 1928, molecular weight inferred, in most quarters, a uniform molecular weight, i.e., all molecules having the same and not a range of statistically averaged values as known today. Meyer's and Mark's explanation was not designed to disagree with high molecular weight but to explain their detection of crystalline - noncrystalline regions and the possibility of a distribution of molecular weights. The misunderstandings grew and the exchange in the literature became polemic.

In an attempt to keep peace, Mark wrote Staudinger in November, 1928, that he was sorry to see that Staudinger was annoyed by Meyer's statements. He went on to say, "I have emphasized the importance of your beautiful work... I prefer not to emphasize the differences, we mean the same thing."

Meyer, however, could not leave Staudinger unanswered, and in a subsequent paper, he restated his views regarding high molecular weight and objected to Staudinger presenting his (Meyer's) views. Staudinger complained of Meyer's failure to cite his work. Meyer said he did often enough. And the exchange continued until a journal editor ended it all by refusing to print their papers on the subject. Caught at that time between two dynamic geniuses, Mark recently recalled:

"Even the champions of the long chain aspect did not agree with each other, as they easily could have done because instead of concentrating on the essential principle, they disagreed on specific details and, at certain occasion, they argued with each other more vigorously than with the defenders of the association theory." (16)

As the debate in Germany raged, other workers in other countries began to work toward a more complete understanding. Wallace Carothers at Du Pont in the U.S.A. began a series of investigations in 1928 which would firmly establish the modern polymer theory. his objective from the beginning was to prepare polymers of known structure through established organic reactions. In the brilliant years before his death in 1937, he studied the preparation and properties of polyesters, polyamides, polyanhydrides, and polychloroprene. As a result of his studies, in which he extended and restated the concepts of Staudinger, Meyer, and Mark with careful reasoning and massive documentation, the high molecular weight concept was accepted without criticism.

REFERENCES

1.   A.V. Lourenco, <u>Ann. Chim. Phys</u>, <u>51</u> 365 (1860); <u>67</u> 293 (1863).

2a.   P.J. Flory, "Principles of Polymer Chemistry," Cornell University Press, Ithaca, 1953, pp. 3-28. The nineteenth century research is summarized quite well in this classic text.

b.   H. Hlasiwetz and J. Haberman, <u>Ann. Chem.</u> <u>Pharm.</u>, <u>159</u> 304 (1871).

3a.   H. Pringsheim, <u>Naturwissenschaften</u>, <u>13</u> 1084 (1925).

b.   M. Bergmann, ibid., <u>13</u> 1084 (1925).

c.   E. Abderhalden, ibid., <u>12</u> 716 (1924).

4a.   H. Crompton, <u>Proc. Chem. Soc.</u>, <u>28</u> 193 (1912), cited in b.

b.   R. Olby, "The Path to the Double Helix," Univ. of Washington Press, Seattle, 1974, pp. 1-37. This book gives a lively recount of the premises of the "association theory" followers as well as some of the events discussed in subsequent sections of this paper.

5.   G. Bouchardat, Comptes Rendus de l'Academie des Sciences, Paris, <u>89</u> 1117 (1879).

6.   W.A. Tilden, <u>Chemical News</u>, <u>46</u> 120 (1882).

7a.   O. Wallach, <u>Annalen der Chemie</u>, <u>239</u> 48 (1887).

b.   W. Wallach, <u>Chemical News</u>, <u>65</u> 265 (1892).

8.   J.H. Gladstone and W. Hibbert, <u>Phil. Mag.</u>, <u>28</u> 38 (1889).

9.   S.S. Pickles, <u>J. Chem. Soc.</u>, <u>97</u> 1085 (1910). This paper is a good, English language, period review of rubber research. Pickles presents the anglophile's side of the smoldering question of that time, "Who first prepared synthetic rubber,

the Englishman Tilden or the Germans F. Hofman and C. Coutelle?"

10.   H. Staudinger, Ber., 53 1073 (1920). This classic review formed the foundation for Staudinger's defense of "macromolecules".

11.   J.R. Katz, Naturwissenschaften, 13 1411 (1925).

12.   A. Fredga, "Les Prix Nobel en 1953," Stockholm, 1954, p. 27.

13.   The interested reader should consult: Ber., 59, 2973, 2982, 3000, 3008, 3019 (1929) for the texts to the Bergmann, Mar, Waldschmidt-Leitz, Pringsheim, and Staudinger presentations.

14.   K.H. Meyer and H. Mark, Ber., 61B 593 (1928).

15.   C. Priesner, "Chemie in unseres Zeit.," 1979 (2), 43.   Priesner gives a well referenced review of the polemics of Staudinger, Mark and Meyer. His book, "H. Staudinger, H. Mark and K. Meyer - Thesen Croesse und Struktur der Makromolekule," Verlag Chemie GmbH, Weinheim, 1981; will soon have an English language edition.

16.   H.F. Mark, and W.O. Milligan ed., "Proceedings of the R.A. Welch Foundation Conferences. X. Polymers," R.A. Welch Foundation, Houston, 1967, pp. 19-58.

SUPPLEMENTAL READING

17.   G.A. Stahl ed., "Polymer Science Overview, A Tribute to Herman F. Mark," American Chemical

Society, Washington, ACS Symposium Series, 175, (1980). This book contains a comprehensive biography of polymer pioneer Herman Mark and, thus, a history of polymer science.

18.   H.F. Mark, "Polymer Chemistry in Europe and America - How it all Began," J. Chem. Educ., 58 (7), 527-534 (1981). This paper is a transcript of some of Mark's recollections.

19.   R.F. Wolf, "Seventy-five Year Stretch," Rubber World, 1964 (10), 64-89. Wolf's review is the most readable account of rubber development.

20.   R.B. Seymour, "History of Polymer Science and Technology," Marcel Dekker, Inc., New York, NY, 1981.

CHAPTER 3

REMEMBERING THE EARLY DAYS OF POLYMER SCIENCE

Herman F. Mark
Polytechnic Institute of New York
Brooklyn, NY 11201

ABSTRACT

Three conferences may be taken as characteristic for the early development of polymer science: The meeting of the Deutscher Naturforscher gesellschaft in Duesseldorf in 1926 where Staudinger successfully upheld his concept of the existence of macromolecules against powerful opposition, the Faraday Society Meeting in Cambridge in 1935 where Carothers presented his classification of addition and condensation polymers, and the First Polymer Conference of the International Union of Pure and Applied Chemistry in Liege in 1947 where polymer science established itself as an accepted and vigorous member of chemical disciplines.

Since the beginning of his existence, man has strongly relied on the use of natural organic polymers for food, clothing, and shelter. When he ate meat, bread, fruit, or vegetables and drank milk, he was feeding on proteins, starch cellulose, and related polymeric materials; when he put on clothing made of fur, leather, wool, flax, and cotton, he used the same natural polymers; and when he protected himself against wind and weather in tents and huts, he constructed these primitive buildings of wood, bamboo, leaves, leather, and fabrics, which also belong to the large family of organic polymers. In addition to the above mentioned types, there are rubber, many resins and bark.

R. B. Seymour (ed.), Pioneers in Polymer Science, 29–39.
© *1989 by Kluwer Academic Publishers.*

Even later, when higher levels of civilization were reached, organic polymers were essential necessities in peace and war. All books in the famous library of ancient Alexandria consisted either of cellulose (paper) or protein (parchment), and books consist of these materials in all libraries of the world up to the present day. All transportation on land and sea throughout many centuries operated on wooden chariots and ships which were put in motion with the aid of ropes and sails made entirely of such cellulosics as flax, hemp, or cotton. The music of all string instruments is produced by the vibrations of wooden, resin-treated boards; and all famous paintings together with many of the most valuable statues consist of cellulose, lignin, and polymerized terpenes on such materials as paper, canvas and wood. Bow and arrows are cellulose, lignin, resin, and protein; catapults and siege towers were made of wood and moved with ropes and - until about 100 years ago - all sea battles were fought with wooden ships which were maneuvered with the aid of cellulosic sails and ropes.

While, in this way, natural organic polymers literally dominated the existence and welfare of all nations, virtually nothing was known about their composition and structure. In each sector - food, clothing, transportation, communication, housing, and art - highly sophisticated craftsmanship developed which was sparked by human intuition, creativity, zeal, and patience and led to accomplishments which deserve the highest admiration of generations to come.

But even when the chemistry of organic compounds became a respectable scientific discipline in the early decades of the last century, the all important helpers of mankind - proteins, cellulosics, starch, and wood - were not in the mainstream of organic chemical research.

Why?

Because somehow they did not seem attractive at that time for a truly scientific study, since they did not respond to the then existing methods for isolation, purification, and analysis. The experimental backbone of organic chemistry in those days was dissolution, fractional precipitation, and crystallization or distillation; it worked and still works with all ordinary organic compounds such as sugar, glycerol, fatty acids, alcohol, and gasoline but fails with cellulose, starch, wool, and silk. These materials cannot be crystallized from solution and cannot be distilled without decomposition.

This fundamental and embarrassing difference between the natural organic materials and the ordinary organic chemicals warned the chemists of the last century that there might be some essential and basic difference between these two classes of substances and that one would have to develop special, new, and improved experimental methods to force the second class into the realm of truly scientific studies.

The breakthrough came in the early decades of this century, mainly through the adoption of physical methods, such as improved optical devices like the ultramicroscope, and the application of the ultracentrifuge, of new viscometers, osmometers, diffusion cells, and most of all, through the systematic application of x-ray and electron diffraction to fibers, membranes, and tissues. A decade of intense research on cellulose, proteins, rubber, and starch wound up with the following fundamental results:

1. All investigated materials consist of very large molecules. Whereas the molecular weights of ordinary organic substances, such as alcohol, soap, gasoline, or sugar range from about 50 to 500, the molecular

weights of the natural organic building materials range from 50,000 to several millions, a fact which earned for them, through Staudinger, the name "giant molecules" or "macromolecules."

2. Most of them have the shape of long flexible chains which are formed by the multifold repetition of a base unit. One often refers to this unit as a "monomer" (monos is the Greek word for "one" and meros is the Greek word for "part") and to the macromolecules themselves as "polymer" (poly is the Greek word for "many").

3. If a sample consisting of regularly built, flexible chains is exposed to mechanical deformation, the individual macromolecules are oriented and show a tendency to form thin, elongated bundles with high internal regularity, which are usually referred to as "crystalline domains" or, simply, as crystallites. Depending upon the nature of the material and the severity of the treatment, a different percentage of the material undergoes crystallization whereas the rest remains in the "amorphous" or "disordered" state so that any given sample - fiber, film, rod, or disk - consists of two phases: amorphous and crystalline.

It was found that the crystalline domains contribute to strength, rigidity, high melting characteristics and resistance against dissolution, whereas the amorphous areas impart softness, elasticity, absorbtivity, and permeability.

As soon as the study of natural polymers had started to establish these ground rules, chemists were strongly tempted to synthesize equivalent systems from simple, available, and inexpensive raw materials. The years 1920 to 1940 brought ever-increasing successful results to:

1.   Provide for more and cheaper basic building units - synthesis of new monomers

2.   Work out efficient equations to describe quantitatively the mechanism of polymerization and polycondensation.

3.   Establish quantitatively the molecular weight and molecular microstructure to arrive at polymer characterization.

4.   Explore the influence of the structural details on the different ultimate properties - molecular engineering.

Learning originally from nature and following up on the established principles, scientists and engineers succeeded in producing a wide variety of polymeric materials which outdo their original native examples in many ways and, in most cases, are much more accessible and less expensive. All this gave a tremendous lift to the important industries of man-made fibers, films, plastics, rubbers, coatings, and adhesives and made everybody's life richer, safer, and more comfortable. Statistics show that about 10 tons of man-made fibers were produced in 1977 with a total value of more than 10 billion dollars; during the same year about 10 tons of synthetic plastics were produced which represent a total value of about 40 billion dollars. Figures of the same order of magnitude hold for synthetic rubbers, coatings, and packaging materials. As a consequence, synthetic organic polymers have become a significant factor in the economy of all industrialized countries in the world.

THREE MILESTONES

For four decades from the 1920's to the 1960's, enormous progress was made in understanding the structure of native polymers and synthesizing artificial counterparts. Instead of giving a detailed chronological account of the events, I shall describe three representative meetings which took place during this period, because they have become classical milestones of modern chemical history.

The first took place in Duesseldorf where the Gesellschaft Deutscher Naturforscher und Artze met in 1926, the second was a meeting of the Faraday Society in Cambridge in 1935, and the third occurred in 1948 in Liege in Belgium and was the Inaugural Session of the High Polymer Commission of IUPAC (The International Union of Pure and Applied Chemistry). I had the privilege of presenting lectures at all three conferences and was chairman of the last.

The meeting in Duesseldorf was arranged in order to confront Staudinger, the protagonist of the macromolecular concept, with several other distinguished scientists who were still reluctant to admit the existence of giant organic molecules and adhered to the idea that many natural substances - cellulose, silk, rubber - consist of small units which are a thousand times larger than those which they themselves were studying in their laboratories. One of them commented: "We are shocked like zoologists would be if they were told that somewhere in Africa an elephant was found which was 1500 feet long and 300 feet high."

To the classical arguments in favor of tightly knit systems of small units, which were presented by prominent scientists such as Karrer, Pringsheim, Bergmann, and Waldschmidt-Leitz, a new one had recently been added;

namely, the fact that the crystallographic basic cells of cellulose, silk, and rubber are so small that they only can contain particles with molecular weights less than one thousand.

According to the then existing classical crystallographic teaching, a molecule cannot be larger than the basic cell. This is where my lecture was supposed to contribute to the discussion by explaining that, under certain conditions, a chemical molecule could well be larger than the basic crystallographic cell and could even be as large as the entire crystallite.

The presentations were followed by rather animated discussions which proved that the differences in opinion were not yet eliminated. At the end, Willstaetter, who presided, summarized his position by saying: "Such enormous organic molecules are not to my personal liking but it appears that we all shall have to become acquainted with them". In fact, the "theory of the small units" lost more and more ground and in the science and engineering the macromolecular character of the principal natural polymers became an accepted fact and immediately encouraged the synthesis and study of any material which could readily be prepared from available monomers which would undergo polymerization in one way or another. So began the first phase of polymer chemistry as a rapidly expanding qualitative scouting for new polymers and their principal properties with the intention of arriving at a general survey of the width and depth of this new branch of organic chemistry.

In 1935 the Faraday Society arranged a meeting on "Phenomena of Polymerization and Polycondensation" in Cambridge, England. It was an international conference, the large size and high level of which proved the enormous progress which the young branch of polymer science had made during the last decade.   There was no question anymore about the existence of macromolecules, for the contributions dealt with the mechanism of polymerization reactions, the determination of molecular weights, the properties of specific polymers as a result of their molecular structure, and potential areas of application. For the first phase the words "observation" and "preparation were valid; for this second phase the words were "measurement" and quantitative characterization." In addition to the old guard - Staudinger, Meyer, Rideal, and myself - a new upcoming generation of excellent scientists presented their work - Melville, Schulz, Houwink, Asbury, and others.   The outstanding figure of this meeting was undoubtedly Wallace Carothers who had come from Wilmington to give an account of the momentous studies which he and his associates had carried out during the last decade.

At the end of the symposium, everyone was convinced that polymer chemistry had grown into a full-scale science with unexpected new vistas for intensification of understanding and expansion of application.   Universities and industrial organizations started to compete with each other to enter the field at any point from fundamental aspect to practical evaluation.   This second, rapidly ascending phase of polymer science was first slowed down and later vigorously accelerated during World War II. The United States and Canada, in particular, were literally running away in basic and applied areas.   Essential fundamental details in polymerization mechanisms including suspension and emulsion polymerization were clarified, and the ground-work for x-ray examination and IR spectroscopy of polymeric systems was firmly established.

After the war the first close contact in science and engineering was made by visiting professors from the

United States to Europe. Such positions were occupied among others by Alfrey, Doty, Mesrobian, Tobolsky, Overberger and myself, who gave several series of lectures in Western Europe right after the war. The interest was understandably very pronounced, and after contact with the president and secretary general of IUPAC in Paris it was decided to organize an International Polymer Symposium in Liege in the summer of 1947; and to initiate a Commission on High Polymers within the Division of Physical Chemistry. There did not, at that time, exist scientific contact with Germany, Italy, Japan, and the USSR and its satellites, but all other countries where work on polymers was done sent strong representative delegations.

At the conference the state of the art was reviewed in a series of comprehensive papers, and the main lines for further progress were in the center of the discussion. Much time was spent to report progress made in England and America to the scientists on the continent. This included synthesis and technology of such polymers as polyethylene, nylon, and polyesters, new methods for structure determination such as x-ray Fourier analysis and polarized infrared; and the beginning of spin resonance and the elaborate use of light-scattering techniques. Present were most scientists who, within the next decade, emerged as leaders of the various branches of polymer science. From England came, among others, Asbury, Bawn, Bernal, Evans, Melville, Rideal, and Thompson. Italy contributed Nasinni and Natta; France sent Champetier, Chapiro, Magat, and Sadron. They had the opportunity to meet Hermans, Houwink, and Staverman from the Netherlands; Errera and Smets from Belgium; Claesson and Ranby from Sweden; Ant-Wuorinen from Finland; W. Kuhn, K.H. Meyer, and Signer from Switzerland.

This conference initiated a vigorous and systematic cooperation of all Western polymer scientists which soon resulted in the essential clarification of all phases of condensation and addition polymerization and including copolymerization, graft, and block polymerization, in an impressive buildup of elastomer and plastics technology, in an almost complete domination of synthetic polymers in the coatings and packaging field. As more polymers with more complicated structures became available, the methods for their quantitative characterization had to be sharpened. All scattering techniques - light, x-rays, electrons, and neutrons - received a thorough overhauling; resonance processes - nuclear and electronic magnetic movements - were build up to admirable precision and, most of all, the study of the chemical structure as well as the state or order (regularity, crystallinity) of polymers by vibrational (infrared and Raman) spectroscopy was intensified and expanded. Today these techniques can be called the most informative and powerful tools for the characterization of even very complex polymeric systems.

In fact, all these methods became more and more necessary because of the enormous number of new polymers, the increasing complexity of their structure, and the wide variety of their application. In order to design a macromolecule for special use, many detailed structural conditions had to be fulfilled which required very precise methods to establish their existence and to control their durability in use.

One of these methods was the vibrational spectroscopy which had its roots in the late 1920's and early 1930's. One of them was the fundamental understanding of molecular vibrations on the basis of quantum mechanics; it was first put in evidence by the absorption of infrared radiation and later also found in the modulations of scattered visible light in the Raman effect. The two methods complemented each other dramatically because of their different response to the selection rules which control the transition probabilities between different vibrational states of a molecular framework. The other root was a gradual and substantial improvement of the experimental techniques, such as stronger and more uniform sources of the primary radiations, higher resolution in the spectroscopic part of the equipment, and, perhaps most of all, more sensitive and reliable receivers.

Another important new and powerful method which greatly advanced our knowledge of the molecular structure of organic systems, including polymers, was the magnetic resonance methods: electron spin and nuclear spin which provide information on the exact relative position of certain atoms - C,N,H - to each other.

And, of course, diffraction experiments - electrons, x-rays and neutrons - some of which already had been important in the earliest days of polymer science are now reaching a new and highly sophisticated level and are an indispensable help in the microcharacterization of polymeric systems.

Thus: the almost limitless number of new experimental techniques for the preparation of new polymeric systems and the availability of powerful tools for their characterization open up a bright future for our field of interest: further growth in the synthesis of new useful materials and ever better understanding of their structure and behavior.

CHAPTER 4

THE "KATZ EFFECT" ON THE RANDOM COIL MODEL FOR POLYMER CHAINS

L.H. Sperling
Materials Research Center
Lehigh Unversity
Bethlehem, PA 18015

ABSTRACT

The x-ray patterns of natural rubber in the relaxed and extended states led J.R. Katz and others to develop a random coil model for polymer chains. The "Katz Effect" which was repeated by H. Mark helped to establish a relationship between mechanical deformation and concomitant molecular events in all macromolecules.

The entropy of a highly extended rodlike molecule is zero but this entropy increases on relaxation because of the existence of random coils. The random coil model has been used to explain both rubber elasticity and many other physical and mechanical phenomena in polymers.

INTRODUCTION

Advances in science and engineering have never been completely uniform, or followed an orderly development. High school and undergraduate students frequently compound the problem by making believe that all science not only had an orderly development, but that there remains little of significance left unknown. The truth is that science advances by fits and jerks, with ideas propounded by individuals who see the world in a different light. Frequently, they face adversity when putting their ideas forward.

R. B. Seymour (ed.), Pioneers in Polymer Science, 41–46.
© *1989 by Kluwer Academic Publishers.*

In 1920, Hermann Staudinger formulated the macromolecular hypothesis: That there was a special class of organic substances of high viscosity which were composed of long chains (1,2). This revolutionary idea was argued throughout important areas of chemistry (3). Finally, it became accepted and formed the most important cornerstone in the development of modern polymer science.

EARLY IDEAS OF POLYMER CHAIN SHAPE

If one accepts the idea of the long chain macromolecule, the next obvious question relates to its conformation or shape in space. This is especially important since the spatial arrangement of the long chains was early thought to be important in determining the physical and mechanical properties of the material. Staudinger himself thought that most amorphous high polymers, such as polystyrene, were rod-shaped, and when in solution, the rods lay parallel to each other.

Rubbery materials were different, however. According to early workers (4), elastomers were coils or spirals in the sense of bedsprings, Staudinger himself (2) described the idea as follows:

"In order to clarify the elasticity of rubber, several investigators have stated that long molecules form spirals, and to be sure the spiral form of the molecules is promoted through the double bonds. By this arrangement, the secondary valences of the double bonds can be satisfied. The elasticity of rubber depends upon the extensibility of such spirals."

THE RANDOM COIL

According to H. Mark (5), the story of the development of the random coil began with the x-ray work of Katz on natural rubber in 1925 (6-9).   Katx studied the x-ray patterns of rubber both in the relaxed state and the extended or stretched state.   In the strethced state, Katz found a characteristic fiber diagram, with many strong and clear diffraction spots, indicating a crystalline material. This contrasted with the diffuse halo found in the relaxed state, indicating that the chains were amorphous under that condition.   The fiber periodicity of the elementary cell was found to be about 9A, which could only accommodate a few isoprene units.   Since the question of how a long chain could fit into a small elementary cell is fundamental to the macromolecular hypothesis, Hauser and Mark repeated the "Katz effect" experiment, and on the basis of improved diagrams and x-rays techniques, established the exact size of the elementary cell (10).   Of course, the answer to the question of the cell size is that the cell actually accommodates the mer, or repeat unit, rather than the whole chain. This story is amplified elsewhere in this book.

The "Katz effect" was particularly important because it was the first experiment to establish the relationship between mechanical deformation and concomitant molecular events in polymers (5).   This led Mark and Valko (11) to carry out stress-strain studies over a wide temperature range together with x-ray studies in order to analyze the phenomenon of rubber reinforcement.   This paper contained the first clear statement that the contraction of rubber was not caused by a decrease in energy but by the decrease in entropy on elongation.

This finding can be explained by assuming that the rubber chains are in the form of flexible coils (12).   These flexible coils have a high conformational entropy, but lose their conformational entropy on being straightened out. The fully extended chain, which is rod-shaped, can have only one conformation, and its entropy is zero.   This concept was extended to all elastic polymers by Meyer, Susich and Valko (13) in 1932. Although thermal motion and free chain rotation are required for rubber elasticity, the idea of the random coil was also adopted for glassy polymers, such as polystyrene.

The main quantitative development began in 1934 with the work of Guth and Mark (14) and Kuhn (15). Guth and Mark chose to study the entropic origin of the rubber elastic forces, while Kuhn was more interested in explaining the high viscosity of polymeric solutions. Using the concept of free rotation of the carbon-carbon bond, Guth and Mark developed the idea of the "random walk" or "random flight" of the polymer chain. This led to the familiar Gaussian statistics of today, and eventually to the famous relationshiop between the end-to-end distance of the chain and the square root of the molecular weight. Three stages in the development of the random coil model have been described by H. Mark (16). Like many great ideas, it apparently occurred to several people nearly simultaneously. It is also clear that Dr. Mark played a central role in the development of the random coil model.

The random coil model has remained essentially the same until today (17), although many mathematical treatments have refined its exact definition. Its main values are two-fold: By all experiments, it appears to be the best model for amorphous polymers, and it is the only model that has been extensively treated mathematically. It is interesting to note that by its very randomness, the random coil model is easier to understand quantitatively and analytically than models introducing modest amounts of order (19).

The random coil model has been supported by many experiments over the years. The most important of these ahs been light-scattering from diulte solutions, and more recently, small-angle neutron scattering from the bulk state. Both of these experiments support the famous relationship between the square root of the molecular weight and the end-to-end distance. The random coil model has been used to explain not only rubber elasticity and dilute solution viscosities, but a host of other physical and mechanical phenomena, such as melt rheology, diffusion, and the equilibrium swelling of crosslinked polymers. Some important reviews include the works of Treloar (18), Boyer (19), Flory (20), and Staverman (21). These general concepts are now discussed in polymer textbooks (22).

REFERENCES

1. H. staudinger, <u>Ber. Dtsch. Chem. Ges.</u>, <u>53</u> 1074 (1920).

2. H. Staudinger, "Die Hochmolekularen Organischen Verbindung," Springer Verlag, Berlin, 1932. Reprinted, 1960.

3. G.A. Stahl, Ed., "Polymer Science Overview - A Tribute to Herman F. Mark," ACS Symp., No. 175, Washington, DC, 1981.

4. F. Kirchhof, <u>Kautschuk</u>, <u>6</u> 31 (1930).

5. H. Mark, Unpublished, 1982.

6. J.R. Katz, <u>Die Naturwissenschaften</u>, <u>13</u> 410 (1925).

7. J.R. Katz, <u>Chem. Ztg.</u>, <u>19</u> 353 (1925).

8. J.R. Katz, <u>Kolloid Z.</u>, <u>36</u> 300 (1925).

9. J.R. Katz, <u>Kolloid Z.</u>, <u>37</u> 19 (1925).

10. E.A. Hauser and H. Mark, <u>Koll. Chem. Beih.</u>, <u>22</u> 63 <u>23</u> 64 (1929).

11. H. Mark and E. Valko, <u>Kautschuk</u>, <u>6</u> 210 (1930).

12. W. Kuhn, <u>Angew. Chem.</u>, <u>49</u> 858 (1936).

13. K.H. Meyer, G.V. Susich, and E. Valko, <u>Kolloid Z.</u>, <u>41</u> 208 (1932).

14. E. Guth and H. Mark, <u>Monatsh. Chem.</u>, <u>65</u> 93 (1934).

15. W. Kuhn, <u>Kolloid Z.</u>, <u>68</u> 2 (1934).

16. H. Mark, Private Communication, May 2nd, 1983.

17. P.J. Flory, "Principles of Polymer Chemistry," Cornell Univ. Press, Ithaca, 1953.

18. L.R.G. Treloar, "The Physics of Rubber Elasticity," 3rd Ed., Oxford, Clarendon Press, 1975.

19. R.F. Boyer, <u>J. Macromol. Sci. Phys.</u>, <u>B12</u> 253 (1976).

20. P.J. Flory, "Statistical Mechanics of Chain Molecules," Wiley, NY, 1969.

21. A.J. Staverman, <u>J. Polym. Sci.</u>, Symp. No. 51, 45 (1975).

22. R.B. Seymour, C.E. Carraher, "Polymer Chemistry: An Introduction," 2nd Ed., Marcel Dekker, New York, NY, (1988).

CHAPTER 5

ANSELM PAYEN
PIONEER IN NATURAL
POLYMERS AND INDUSTRIAL CHEMISTRY

Charles H. Fisher
Department of Chemistry
Roanoke College
Salem, VA 24153

ABSTRACT

Anselm Payen, an early pioneer in polysaccharide, industrial, and agricultural chemistry, was the first to separate cellulose from wood and plants. He showed that celluloses and starches from various sources have the same composition ($C_6H_{10}O_5$) and are isomeric with each other. Payen observed (with Pensozok) that starch is hydrolyzed by a substance in malt, which they called diastase. Other polymers investigated by Payen include dextrins, lignin, rubber, and gutta percha; these and other investigations were described in some 200 papers and ten books. As owner and manager of plants that manufactured sal ammoniac, sulfuric acid, hydrochloric acid, borax, refined sulfur, soda, gelatin, charcoal, and sucrose from sugar beets, Payen developed improved processes and introduced new commercial operations. Payen received prestigious honors during his life and two approximately one hundred years later. In 1962 the Anselm Payen Award was established by the ACS Cellulose Paper and Textile Division. In 1968 Payen was named one of the greatest scientists of all time when his biography was included in the World's Who's Who in Science. Born in Paris in 1795 and deceased in 1871, Payen was buried in Grenelle near Paris.

R. B. Seymour (ed.), Pioneers in Polymer Science, 47–61.
© *1989 by Kluwer Academic Publishers.*

PAYEN'S INVESTIGATIONS OF NATURAL POLYMERS

Although Frenchman Anselm Payen investigated several natural polymers and polymer-rich materials (3, 11, 14, 23, 40, 41, 43, 44), he may be remembered most for his pioneering research on cellulose, the most abundant polymer and organic material in the world. Cellulose, manufactured by nature at the incredible rate of perhaps 200 billion tons per year worldwide, is the principal component of materials (cotton, linen, jute, wood, etc.) that have been known and used since earliest times (21, 45).

Nevertheless, many thousands - perhaps millions - of years passed before cellulose was discovered and investigated as a distinct and separate chemical entity. The story of another natural polymer, starch, is similar. Starch, a major component of many foods and feed, sustained humans and other animals for thousands of years before being identified and treated as a chemical substance.

It was as late as the 19th century that the brilliant researches of Anselm Payen played a major role in bringing about a better understanding of these polysaccharides.

Payen's classic work on cellulose was described in his articles entitled "Study of the Composition of the Natural Tissues of Plants and of Lignin" and "Concerning a Means for Isolating the Elemental Tissue of Wood" which were published in 1838 in Comptes Rendus. Payen treated plant tissues, cotton linters, root tips, and various woods with chemicals, including nitric acid in some instances, followed by extraction with ammonia or alkali, water, and solvents, to obtain a fibrous substance, named cellulose. He demonstrated that this "elemental" product from plant sand wood has the composition of starch and is apparently isomeric with it. He showed that, regardless of origin, cellulose is the same and has the chemical composition, represented by the formula $C_6H_{10}O_5$.

Payen observed that, to isolate cellulose, it is necessary to remove substances having a higher percentage of carbon than cellulose. Payen called these substances "incrusting materials." These materials were later (1857) designated by Schulze as lignin, a term used previously by Candolle (41).

A Committee of the French Academy evaluated Payen's work and reported in 1839: "...the observation by M. Payen shows that lignin belongs to a different class than starch sugars..." "He has performed an exact and successful separation of the two organic elements of wood." "In fact, in wood, there is the basic tissue, which is isomeric with starch, which we will call cellulose, and also a substance which fills the cells and which constitutes the true ligneous matter" (44).

In short, Payen showed that cellulose and lignin can be separated. Nitric acid dissolves the lignin, leaving cellulose, whereas sulfuric acid dissolved the cellulose and leaves lignin as a residue (12, 13).

Payen has been given credit (42) for suggesting the name cellulose, but apparently this honor may belong to the above-mentioned Committee, consisting of M. Brogniart (founder of modern paleobotany), M. Pelouze, and M. Dumas (44).

Payen studied starch extensively and clearly recognized starch as a reserve carbohydrate in plants. He showed that starches from different sources have the same composition. The claim by Debus that Payen discovered dextrins in 1836 and pectin in 1824 has been disputed (11).

In 1833, Payen (with Persoz) found that the transformation of starch into dextrin and sugar by malt (discovered by Kirchhoff) was due to a substance, diastase, extractable with water from germinated barley. They purified the material by repeated precipitation with ethanol and found that the activity was destroyed at 100°C (3, 39).

Diastase is an example of a biological catalyst, which was called an enzyme by Kuhne a half a century later. Diastase was the first enzyme to be isolated in concentrated form; its name started the practice of using the suffix "-ase" in naming enzymes (3).

Payen's status as an early pioneer in giant molecules is strongly supported by his investigations and numerous publications on polymers and polymer-containing materials, including cellulose, starch, rubber, gutta percha, gelatin, dextrins, wood, plants, and meat (44).

PAYEN: TECHNOLOGIST   AND   CHEMICAL MANUFACTURER

Versatile Payen was both a curious scientist and practical technologist. Even when searching for new scientific information, Payen usually looked for possible practical applications. His inherited wealth, which included chemical manufacturing plants, facilitated his efforts in both science and technology (41).

In 1792 Jean Payen, the father, established a factory at Grenelle, France for the production of sal ammoniac (from the destructive distillation of bones and other animal byproducts), sulfuric acid, hydrochloric acid, borax, sulfur, soda, and gelatin. He established a factory in Vaugirard for the production of sugar (sucrose) from sugar beets (41).

In 1815, at the age of twenty, Anselm Payen was made manager by his father of a plant that imported crude borax from the East and refined it for sale as a commercial product. Payen made his first major contribution to industrial technology by developing a low-cost method of preparing borax from boric acid and soda. In 1820 he began marketing the synthetic product at one-third the price of the refined natural borax, thereby establishing a new industry in France.

In 1820, following his father's death, the 25-year old Payen inherited the full responsibility for managing several factories, one of which manufactured beet sugar.

In 1822, Payen published a paper ("Theory of the Action of Animal Charcoal and Its Application to the Refining of Sugar") describing his thorough study of the decolorizing properties of animal charcoal. The paper pointed out the decolorizing activity was due to the shape and state of aggregation of carbon. In addition, the paper showed that charcoal could remove certain salts from solution, thereby facilitating crystallization of the sugar. As a result of Payen's work, a waste product became the basis of a new industry, namely, the production and use of animal charcoal.

Payen developed a "decolorimeter" for evaluating the decolorizing ability of various lots of charcoal. Charcoals of the general types developed by Payen have been utilized down through the years for important uses, including gas masks in World War I and various refining and purification procedures.

Payen's book (1826) entitled, "Treatise on the Potato" described the preparation of various materials from the potato, including foods, feeds, starch, syrup, sugar, and alcohol. The Central Agricultural Society of France gave Payen a gold medal in 1823 in recognition of this work.

In the early 1800's, economical methods were needed for disposing of or utilizing the carcasses of domestic animals. The problem was grave enough that the Central Agricultural Society of France in 1825 offered a first prize of 1,000 francs for the best method of handling and using the carcasses. In 1830, the first prize went to Payen for his work and suggestions, which were described in a 132-page memoir.

Nitrogen was an important component of some of the materials used or produced in Payen's chemical plants. Payen developed a method for determining nitrogen, which consisted in heating the sample to red heat and collecting the volatile products in dilute sulfuric acid. The process was modified by Will and Varrentrapp and was superceded later by the well-known Kjeldahl method (41,44).

LIFE AND CAREER OF ANSELM PAYEN

Anselm Payen, born in Paris, France, on January 6, 1795 (deceased in the same city, May 12, 1817), was the son of Jean Baptiste Pierre Payen and Marie Francois Jeanson de Courtenay. His father was educated at the College de Navarre in Paris, where he distinguished himself, particularly in philosophy and the sciences. He studies law also and was, for a time, assistant to the procurator of the King for the city of Paris (41, 44).

Anselm Payen studied first under the direction of his distinguished father and then at Ecole Polytechnique in Paris, where he studied chemistry under Vauqueln (1762-

1829), physics under Fourcroy (1755-1809), and mathematics under Tremery. (Farrar claims Payne did not attend Ecole Polytechnique (14)).

Some of Payen's early research was done to support and improve the chemical manufacturing business which he inherited from his parents. In 1835, at the age of forty, Payen left active participation in his manufacturing enterprises to become Professor of Industrial and Agricultural Chemistry at the Ecole Centrale des Arts et Manufactures. Four years later, he accepted a second academic position as Professor of Applied Chemistry at the Conservatoire des Arts et Metires. He held both academic posts until his death in 1871 (41, 44).

During his approximately fifty-year career in science and technology, Payen published about 200 papers in the major French scientific journals. These described research on dextrin, sugar, bitumen, lignin, cellulose, starch, the enzyme diastase, charcoal, potato, waste utilization, analytical methods, plant and animal nutrition, rubber, gutta percha, manure, water supply, and phylloxera.

Payen, a member of many scientific societies, was elected a member of the Central Agricultural Society of France in 1833 and was its' secretary for 26 years. He was also a member of the French Academy of Sciences, the Academy of Medicine, the Society of the Advancement of National Industry, the Seine Horticultural Society, and the Council of Hygiene and Public Health.

Payen served the French Government as a member of various commissions. Louis Phillippe made him an officer of the Legion of Honor in 1847, and Napoleon III elevated him to commander in 1863.

In 1821, Payen married Zelie Charlotte Melanie Thomas; five children were born to this happy union. Four of the children, however, died in childhood and only one, a daughter, survived Payen.

Intensely patriotic, Payen served for forty years as Commander of the battalion of the National Guard in Grenelle, a suburb of Paris in which he lived.

In old age, Payen witnessed the disastrous end of the rule of Napoleon III in the flames of the Franco-Prussian War. In spite of this age of 76 years he refused to leave Paris as the Prussians approached.

His last days were spent appropriately as a French patriot and scientist using his knowledge to benefit humanity. In 1871 when Paris was under siege by the Prussians, Payen struggled to find sources of food for the starving Parisians. On May 9 when civil war was raging in the streets of Paris, the proud old man attended his last scientific meeting at the French Academy of Medicine. Fatally stricken during a session, Payen died within three days on May 12, 1871. He was buried in the Cemetery at Grenelle.

A truly remarkable man, Payen made major scientific and technological contributions in several different fields. He was easily a peer of the other great scientists of that period (44).

Payen received many honors, including:

1826- Gold Medal of the French Agricultural Society.

1828-Knight of the Legion of Honor by Charles X of Sweden.

1833-Secretary of the Agricultural Society of France for 26 years.

1847-Officer of the Legion of Honor by Louis Phillippe of France.

1863-Commander of the Legion of Honor by Napoleon III.

Nearly 100 years later, Payen received further recognition. In 1962, the American Chemical Society's Cellulose, Paper & Textile Division established the Anselm Payen Award to honor and encourage outstanding professional contributions to the science and chemical technology of cellulose and its allied products.

In 1968 Payen was named as one of the great scientists of all history by inclusion of his biography in the World's Who's Who in Science (11).

REFERENCES

1.   Alfrey, Turner, in "Polymer Science Overview. A Tribute to Herman F. Mark," G.A. Stahl, Editor American Chemical Society, Washington, DC, 1981.

2.   Anderson, R.A., and Watson, S.A., in "Handbook of Processing and Utilization in Agriculture," Vol. II, Part 1, I.A. Wolff, Editor, CRC Press, Inc., Boca Raton, FL, 1982.

3.   Asimov, Isaac, "Biographical Encyclopedia of Science and Technology," Second Edition, Doubleday & Co., Inc., Garden City, NY, 1982.

4.   Billmeyer, F.W., Jr., "Textbook of Polymer Chemistry," Interscience Publishers, NY, 1957.

5.	Bolker, H.I., Natural & Synthetic Polymers, Marcel Dekker, Inc., NY, 1974.

6.	Brogniart, A., Pelouze, M., and Dumas, A.B., Comptes Rendus, 8, 51 (1839).

7.	Brown, N.C., "Forest Products: Their Manufacture and Use," John Wiley & Sons, Inc., NY, 1927.

8.	Carruth, G., editor, "Encyclopedia of Facts and Dates," 4th Ed., Thomas Y. Crowell Co., NY, 1966.

9.	Craver, J.K., and Tess, R.W., "Applied Polymer Science," American Chemical Society, Washington, DC, 1975.

10.	Dean, G.R., pages 961-969 in "Kirk-Othmer's Encyclopedia of Chemical Science & Technology," First Edition, Vol. 4 (1949), Interscience Publishers, NY.

11.	Debus, A.G., Editor, World's Who's Who in Science, p. 1319, Marquis Who's Who, Inc., Chicago, IL, 1968.

12.	Farber, E., "The Evolution of Chemistry," Ronald Press Co., NY, 1952.

13.	Farber, E., "Great Chemists," Interscience Publishers, NY, 1961.

14.	Farrar, W.V., Anselm Payen, p. 435, in "Dictionary of Scientific Biography," C.C. Gillispie, editor, Charles Scribner's Sons, NY, 1974.

15.	Fieser, L.F., and Fieser, Mary, "Organic Chemistry," D.C. Heath & Co., Boston, MA, 1944.

16.   Fisher, C.H., in "History of Polymer Science and Technology," R.B. Seymour, editor, Marcel Dekker, Inc., NY, 1982.

17.   Fisher, C.H., and Chen, J.C.P., in "Handbook of Processing and Utilization in Agriculture," Vol. II, Part 1, I.A. Wolff, editor, CRC Press, Inc., Boca Raton, FL, 1982.

18.   Fisher, H.L., "Organic Chemistry" in Chemistry, Key to Better Living, Diamond Jubilee Volume, American Chemical Society, Washington, D.C., 1951.

19.   Flory, P.J., "Principles of Polymer Chemistry," Cornell University Press, Ithaca, NY, 1953.

20.   Friedel, R., "Pioneer Plastic: Celluloid," University of Wisconsin Press, Madison, WI, 1983.

21.   Goldstein, I.S., "Organic Chemicals from Biomass," CRC Press, Inc., Boca Raton, FL, 1981.

22.   Gadd, P.A., Address given in 1964 in Finland (English translation by Johan Bjorksten, The Chemist, April 1964, pp. 147-156.

23.   Gillispie, C.C., Editor, "Dictionary of Scientific Biography," Vol. 10, Charles Scribner & Sons, NY, 1974.

24.   Haynes, W., "Cellulose, The Chemical that Grows," Doubleday & Co., Garden City, NY, 1953.

25.   Hergert, H.L. and Daul, GlC., Chapter 1 in "Solvent Spun Rayon, Modified Cellulose Fibers, and Derivatives," A.F. Turbak, Editor, American Chemical Society, Washington, DC, 1977.

26.   Hochheiser, S., <u>Science 200</u>, 818-819 (1983).

27.   Horne, W.D., Cantor, S.M., and Liggett, R.W., "Sugar Chemistry," pp. 127-131 in "Chemistry, Key to Better Living," Diamond Jubilee Volume, American Chemical Society, Washington, D.C., 1951.

28.   Houwink, R., "Elastomers and Plastomers," Elsevier Publishing Co., Inc., New York, 1949.

29.   Ihde, A.J., "The Development of Modern Chemistry," Harper and Row, New York, 1964.

30.   Kerr, R.W., pp. 764-778 in "Kirk-Othmer's Encyclopedia of Technology," 1st Ed., Vol. 12 (1954), Interscience Publishers, New York.

31.   Lane, M., and McCombes, J.A., Chapter 12 in "Solvent Spun Rayon, Modified Cellulose Fibers, and Derivatives," A.F. Turbak, Editor, American Chemical Society, Washington, D.C., 1977.

32.   Leicester, H.M., "Historical Background of Chemistry," John Wiley & Sons, Inc., New York, 1956.

33.   Marchessault, R.H., Chapter 4 in "Milton Harris: Chemist, Innovator, and Entrepreneur," American Chemical Society, Washington, D.C., 1982.

34.   McCarthy, K., <u>Exxon USA</u>, Second Quarter, 1983, p. 25.

35.   Meyer, K.H., "Natural and Synthetic High Polymers," Second Edition, Interscience Publishers, Inc., New York, 1950.

36. Mitchell, R.L., and Daul, G.C., pp. 810-847, Vol. II, "Encyclopedia of Polymer Science & Technology," H.F. Mark and N.G. Gaylord, Editors, Interscience Publishers, Inc., New York, 1969.

37. Moore, F.J., and Hall, W.T., "A History of Chemistry," McGraw-Hill Book Co., New York, 1939.

38. Morgan, P.W., chapter in "History of Polymer Science & Technology," R.B. Seymour, Editor, Marcel Dekker, Inc., New York, 1982.

39. Ostwald, W., Nobel Lecture, 1909, in "Nobel Lectures, Chemistry 1901-1921," Elsevier Publishing Co., New York, 1966.

40. Phillips, Max, J. Wash. Acad. Sci., 30, 65 (1940).

41. Phillips, Max, Anselm Payen, pp. 497-504, in "Great Chemists," Edward Forbes, Editor, Interscience Publishers, Inc., New York, 1961.

42. Purves, C.B., Cellulose Historical Survey, pp. 29-53, in "Cellulose and Cellulose Derivatives," O.H. Emil, Editor, Part I, Interscience Publishers, Inc., New York, 1954.

43. Reid, J.D. and Dryden, E.C., Textile Colorist, 62, 43 (1940).

44. Schwenker, R.F., Jr., "The Real Payen," talk presented at the 168th American Chemical Society Meeting in Atlantic City, NJ, 1974.

45. Seymour, R.B., and Carraher, C.E., Jr., "Polymer Chemistry," 2nd Ed., Marcel Dekker, Inc., New York, 1988.

46. Seymour, R.B., Editor, "History of Polymer Science & Technology," Marcel Dekker, Inc., New York, 1982.

47. Smith, A.R., Chapter 10 in "Chemical Aftertreatment of Textiles," H. Mark, N.S. Wooding, and S.M. Atlas, Editors, Wiley-Interscience, New York, 1971.

48. Stahl, G.A., Editor, "Polymer Science Overview: A Tribute to Herman F. Mark," American Chemical Society, Washington, D.C., 1981.

49. "Thorpe's Dictionary of Applied Chemistry," 4th Ed., Vol. IX, (1949), Longmans, Green, and Co., New York.

50. Wakeman, R.L., "Chemistry of Commercial Plastics," Reinhold Publishing Corp., New York, 1947.

51. Whistler, R.L., and Zysk, J.R., pp. 535-555 in "Kirk-Othmer's Encyclopedia of Chemical Technology," Third Edition, Vol. 4 (1978), Interscience Publishers, New York.

52. Whistler, R.L., and Daniel, J.R., pp. 492-507 in "Kirk-Othmer's Encyclopedia of Chemical Technology," Third Edition, Vol. 21, (1983). John Wiley and Sons, New York.

53. Wint, R.F., and Vanderslice, C.W., Chapter 57 in "Applied Polymer Science," J.K. Craver and R.W. Tess, Editors, American Chemical Society, Washington, D.C., 1975.

54. Wise, L.E., The Paper Industry, April 1939, p. 38.

55.   Wolfrom, M.L., Chapter 16 in "Organic Chemistry,"
      H. Gilman, Editor, Vol. II, John Wiley and Sons,
      Inc., New York, 1938.

56.   Zollinger, H., Chapter 10 in "Cellulose and Fiber
      Science Developments," J.C. Arthur, Jr., Editor,
      American Chemical Society, Washington, D.C., 1977.

57.   Powers, P.O., "Synthetic Resins & Rubbers," John
      Wiley and Sons, Inc., New York, 1943.

CHAPTER 6

EMIL FISCHER
PIONEER IN MONOMER AND POLYMER SCIENCE

Charles H. Fisher
Chemistry Department
Roanoke College
Salem, VA 24153

ABSTRACT

Emil Fischer, recipient of the Nobel Prize in 1902 and one of chemistry's alltime greats, invested much of his outstanding career in pioneering the science of monomers, polymers, polymerization, and depolymerization. His remarkable genius becomes even more evident when it is recalled that he lived (1852-1919) and worked with complex materials long before the development of modern instruments and techniques. The monomers of his investigations included formaldehyde, hydroxyaldehydes, sugars, amino acids, and hydroxybenzoic acids. The monomeric aldehydes were converted into carbohydrates of increased molecular weight; the synthesis of starch and cellulose was contemplated. Various amino acids, some prepared by the hydrolysis of proteins, were polymerized to polypeptides. Tannins were hydrolyzed to hydroxybenzoic acids, e.g., gallic acid; these were converted into polyesters of sugars.

INTRODUCTION

Emil Fischer, recipient of the Nobel Prize and many other honors, has long been acclaimed appropriately as one of chemistry's all-time greats (1-20). His genius was displayed in several areas of research, and he is

63

R. B. Seymour (ed.), Pioneers in Polymer Science, 63–80.
© 1989 by Kluwer Academic Publishers.

remembered as a great organic chemist. But to characterize Fischer exclusively as a great organic chemist would overlook his remarkable versatility, wide interests, capabilities as an educator and research administrator, and his important achievements in research on industrial products, e.g., dyes and drugs, and monomers and polymers. The principal dyes, studies by Fischer, were members of the phthalein and rosaniline families. Some of the barbituates and purines prepared by Fischer became commercially important.

Monosaccharides, polysaccharides, amino acids, polypeptides, proteins, and depsides were investigated extensively by Fischer. His monomers related to the polysaccharides included formaldehyde, hydroxyacetaldehyde, several other polyhydroxyaldehydes and the sugars. Fischer's monomers related to polypeptides and proteins were the various amino acids and oligopeptides. Phenol carboxylic acids (or hydroxybenzoic acids) were the monomers used to prepare depsides.

Mother Nature, generous in supplying an abundance of giant molecules, has not been as generous in the science of her polymers which are complex and present many research difficulties. The difficulties were particularly severe before and during the life (1852-1919) of Emil Fischer because modern instruments and techniques had not been developed. Nevertheless, Fischer had both the courage to undertake pioneering research on nature's complex macromolecules and the competence needed for success. His creative research comprises an important milestone in man's search for an understanding of polymers.

In connection with his monomer-polymer research, Fischer pointed out that the polysaccharides, e.g., starch and cellulose, can be hydrolyzed to glucose and, conversely, the monosaccharides such as glucose, can be transformed into polysaccharides. Fischer stated (21) that there is no fundamental difference between glucosides and polysaccharides, the latter being nothing more or less than the glucosides of the sugars themselves. Following this concept, Fischer prepared dextrin-like substances and synthetic disaccharides and he contemplated the synthesis of the still higher polysaccharides. Similarly, Fischer hydrolyzed proteins into amino acids, and polymerized these amino acids to obtain polypeptides. One of these polypeptides contained nineteen amino acid units.

During many years of intense studies, Fischer never postulated any other structure for the synthetic polypeptides or natural proteins except linear chains consisting of covalently linked amino acids connected to each other by the -CO-NH- linkages, now known to occur in all amides and peptides. In 1906, he proposed that there was an uninterrupted continuous line between the dimeric and trimeric amino acids and the native proteins (22). Similarly, Fischer envisioned the polysaccharides, including starch and cellulose, as being polymers of glucose.

Herman F. Mark, a more recent great polymer scientist, expressed his appreciation of Fischer's polymer research as follows (11-14):

"The groundwork for the organic chemistry of polymers, or macromolecules, was laid around 1905 in the Institute of Emil Fischer in Berlin. His work on sugars and amino acids clarified, in a complete manner, the composition, structure, and stereochemistry of these substances and opened the way to a step by step synthesis of progressively larger and larger molecular species. Fischer himself remained strictly in the domain of classical organic chemistry, of which he was the unsurpassed master, and reached only the lower limits of polymer chemistry (molecular weights between 1000 and 1500), but his coworkers pioneered in all fields of true polymer research. Freudenberg, Helferich, and Hess pioneered in the field of polysaccharides; Leuchs and Bergmann in the domain of polypeptides; and Harries and Pummerer in the area of polyhydrocarbons, particularly rubber."

EMIL FISCHER'S LIFE

(Hermann) Emil Fischer was born on October 9, 1952, at Euskirchen, in the Cologne district of Germany. The family was Protestant and had been in the Rhineland since the end of the seventeenth century. Fischer was the son of Laurenz Fischer, a successful merchant, and Julie Poensgen Fischer. The family consisted of the parents, one son, Emil, and five daughters. The son had a pleasant youth; his playmates were mostly his numerous cousins. The friendship with one cousin, Otto Philipp Fischer, developed into a cooperative relationship in university studies and research.

Emil studied three years with a private tutor and then attended the local public school for four years. This was followed by two years of schooling at Wetzler and two more at the gymnasium in Bonn. He passed his final examination in 1869 at the latter institution with great distinction.

There was disagreement about choosing a career. His father wanted to maintain the family tradition and to train Emil to become a successor in business. Emil preferred natural science, mathematics, and especially physics. Emil's trial period in business was a failure. His father admitted defeat and consented to a university education. It is reported that the father said: "The boy is too stupid to be a business man, so he had better become a student." (When his father died years later at the age of 95, Fischer expressed regret "... that he did not live to see his impractical son receive the Nobel Prize in Chemistry.")

The start of Emil's higher education had to be delayed because the eighteen-year-old youth had to recover from gastric catarrh before he could leave home. The father sent Emil to the University of Bonn in 1871 to study chemistry. At Bonn he attended the lectures of Kekule, Engelbach, and Zincke. Emil still had a strong interest in physics, and only the persistent persuasion of his cousin and fellow student, Otto Fischer (1852-1932) prevented him from deserting chemistry.

Emil and his cousin transferred to Strasbourg in the fall of 1872. Studying under Rose, Fischer became acquainted with Bunsen's methods for the analysis of water. This experience proved useful when the young man did analytical work for the town of Colmar.

While studying organic chemistry, Emil was attracted to Adolf von Baeyer, a young professor and future discoverer of the synthesis of indigo. Baeyer was influential in causing Emil to develop enthusiasm for chemistry and suggested the thesis topic of phthalein dyes. Emil's research was successful and the doctorate was conferred in 1874. In the same year, Fischer was appointed assistant instructor at Strasbourg. While at Strasbourg, he discovered phenylhydrazine and demonstrated its relation to hydrazobenzene and the sulfonic acid described by Strecker and Romer.

Emil devoted most of his time to chemistry during his seven semesters at Strasbourg. However, he also studied physics under August Kundt and Wilhelm Rontgen and mineralogy under Paul Groth.

In 1875, Baeyer was asked to succeed Liebig at the University of Munich. He persuaded his young assistant to accompany him to the Bavarian capital. His cousin, Otto, also transferred to Munich. In the fall of 1875, Fischer accepted an assistantship in the organic division of the chemistry department. In 1878, he qualified as Privatdozent at Munich. He was appointed Associate Professor of Analytical Chemistry there in 1879. In the same year, he declined the offer of a chair of chemistry at Aix-la-Chapelle. At the age of twenty-seven, Fischer declined his first call to a full professorship at the Technical University of Aachen. He preferred to work with the inspiring group at Munich.

However, in 1882, he accepted an appointment as full professor to the chair in chemistry at Erlangen. In 1883, the Badische Anilin und Soda-Fabrik asked Fischer to take over the direction of its scientific laboratory. The offer was tremendously attractive because of the higher salary and the excellent facilities available for research. Nevertheless, Fischer refused to sacrifice the freedom of his academic surroundings. His inherited independent financial status, presumable, made the declination much easier.

The latter part of his Erlangen period was darkened by a serious illness. An obstinate chronic catarrh, which also attacked the intestinal tract, forced him to take a year's leave of absence. For the second time in his life, his serious ailment warned Fischer not to overexert himself. From that time on, he lived more carefully and worried excessively about simple colds. It was his health that influenced him to decline the flattering offer from the

Federal Technical University of Zurich, where he would have succeeded Victor Meyer.

Fischer considered his seven years of professorship at Wurzburg (1885-1892) as being the happiest of his career. It was at Wurzburg in 1888 that Fischer married Agnes Gerlach and became the father of three sons. This attractive and kind woman gave understanding care to her husband. Unfortunately, she died of a middle-ear infection only seven years after their marriage.

Fischer, who abhorred all intolerance of religious and political faiths, was pleased with the tolerance that featured life in Wurzburg. Although a Protestant (of Jewish ancestry), Fischer was on a friendly terms at Wurzburg with a number of the representatives of the Catholic theology. By order of the Bishop of Speyer, twenty-five Catholic theologians regularly attended Fischer's lectures, partly because many of their parishioners were employed in the chemical firm of Badische Anilin Und Soda-Fabrik.

At a special meeting of the German Chemical Society on June 23, 1890, Fischer reviewed comprehensively the status of carbohydrate research. It must have been an unusually attractive lecture. The normally critical Carl Harries wrote:

"I have never heard a better lecture with respect to form and content, filled with enthusiasm and genuine moderation; in it the truly great investigator came clearly into view. Emil Fischer became for us the yardstick by which to measure all other personalities."

Following the death of A.W. Hoffmann in the spring of 1892, the University of Berlin invited Emil Fischer to be

the successor. It was not easy for Fischer to leave
Wurzburg. Nevertheless, in 1892 and at the age of forty,
he went to Berlin to assume the most important chair of
chemistry in Germany. During his career at Berlin, which
lasted until has death in 1919, Fischer received the Nobel
Prize in 1902 for his work on sugars and purines.
Nevertheless, it has been said that his work after 1902 on
proteins was more deserving of recognition than his work
before 1902.

In 1907, the Faraday medal of the English Chemical
Society was presented to Fischer. This required a trip to
England to deliver the Faraday lecture. A previous
invitation to give the Faraday Lecture in 1895 had been
declined by Fischer because of poor health.

Fischer's laboratory at Berlin became the organic
chemistry center of the world. Many students came to
Berlin to study under the famous master and important
visitors from many parts of the world came to confer with
Fischer and his able colleagues.

His position in Berlin brought him many
responsibilities and duties. He served several times as the
president and vice-president of the German Chemical
Society. He was a member of the Prussian Academy of
Sciences. He played an active and fruitful role in founding
the pure research laboratories under the auspices of the
Kaiser Wilhelm Gesellschaft. One of the first of these was
the Kaiser Wilhelm Institut fur Chemie, built at Dahlem
in 1911 on the outskirts of Berlin. The founding of the
Kaiser Wilhelm Institut fur Kohlenforschum at Muhheim
was completed in 1914 just before the outbreak of World
War I.

During World War I, Fischer held a high position
under the Hohenzollens. He supported his native

Germany by organizing German chemical resources; recommending conservation of food supplies; advising the Government, especially on matters such as synthetic food, e.g., ester margarine from fatty acids; proposing camphor substitutes and new sources of glycerol; increasing the supply of ammonia from coke ovens; and encouraging the synthetic nitric acid industry. Although Fischer did much for Germany during the war, he never regarded the conflict as being waged in the best interests of the German people. This view jeopardized his social position. After the war, Fischer helped reorganize the teaching of chemistry and reestablish research facilities.

That Fischer was unhappy about WWI is indicated by his letters sent to Theodore W. Richards of Harvard University (6). Excerpt from December 23, 1911 letter: "Here, day by day, one is more afraid of a warlike development, which in my opinion would be a real disaster for Europe." Excerpts from a November 16, 1914 letter: "This (delay of five weeks in delivery of letter) is the best evidence of how much the horrible war, under which half of the world is suffering, interferes with transportation. You know that I foresaw the disaster. But the reality, by far, surpasses all we imagined then." "... some families already are experiencing great grief through heavy losses in the battlefields, including our academic circles. Nernst, Sachu and the chemist Professor G. Krawmer each lost one son. The son of von Planck has been wounded and is a prisoner in France. Some of my associates have died in combat and we are anticipating new losses daily."

World War I was a heavy blow for Emil Fischer. His research activities were curtailed and conducted only with great difficulty. He not only suffered the hardships of the war and the German defeat, but also the loss of two of his three sons. One was killed in World War I, and another committed suicide at the age of twenty-five, as the result of compulsory military training. The third son, Hermann Otto Laurenz Fischer, had a distinguishing career in chemistry, and was Professor of Biochemistry at the University of California, Berkeley prior to his death in 1960.

Fischer's genius achievements were recognized and applauded both during his life and after his death in 1919. He was invited to accept important positions in prestigious institutions. He accepted such offers from the universities of Munich, Erlangen, Wurzburg, and Berlin. He declined similar invitations from the Federal Technical University of Zurich, the University of Aix-la-Chapelle, and Badische Anilin Und Soda-Fabrik. The invitation to become research director of the last-mentioned organization shows that Emil Fischer's abilities were appreciated not only by academia but also by industry.

The honors received by Emil Fischer include the Nobel Prize (1902), the Prussian Order of Merit, and the Maxmillian Order for Arts & Sciences. He was made a Prussian Geheimrat (Excellenz). He received honorary doctorates from the Universities of Christiana (Cambridge), Manchester, and Brussels. He received the Davy Medal of the Royal Society of 1890, and was elected foreign member in 1899. he was honored with the Faraday Medal of the English Chemical Society in 1907 and with honorary membership in the American Chemical Society (23). In 1968 Fischer was honored again as one of the greatest scientists in all history when his biography was included in the World's Who's Who in Science (2).

Fischer's researches on carbohydrates, purines, and proteins were of such enormous importance that, at the repeated requests of the scientific community, they were published in three bulky volumes, covering the periods 1882-1906 (purines), 1884-1908 (carbohydrates and enzymes), and 1899-1906 (amino acids, polypeptides, and proteins).

Following Emil Fischer's death in 1919, the German Chemical Society established the Emil Fischer Memorial Medal. In 1952, Fischer's oldest son dedicated the Emil Fischer Library, containing his father's books and reference compounds, at the University of California at Berkeley.

Many reviews, some lengthy and detailed, of Emil Fischer's life and works have been published in both German and English. Burckhardt Helferich, an assistant to Fischer and subsequently a professor of chemistry at Frankfurt-Main, Greifsward, Leipzig, and Bonn, ended his review of Fischer's life and work with the following words (20):

> "Scientific accomplishments, planned and carried out with farseeing objectives, but firmly rooted in carefully executed small studies, an unswervable love of truth, which invariably submitted to the experimental findings no matter how seductive the theory, a keen understanding and an artistic intuition in minor as well as major occasions, reveal the great chemist in Emil Fischer. His comprehension of other problems in chemistry and the natural sciences and his entire personality make him one of the truly great German scientists. As stated by Richard Willstatter and echoed by all who knew Emil Fischer and his work: He was the unmatched classicist, master of organic-chemical investigation with regard to analysis and synthesis, as a personality a princely man."

EMIL FISCHER'S RESEARCH

Mono and Polysaccharides. At the beginning of Emil Fischer's carbohydrate research in 1884, there were four known monosaccharides: glucose, galactose, fructose, and sorbose, all having the formula $C_6H_{12}O_6$. There were three known disaccharides: sucrose, maltose, and lactose. Fischer elaborated the complex structure and chemistry of the carbohydrates, synthesized many of them, and established the configurations of the sixteen possible stereoisomers of glucose.

Fischer used the Kiliani method (involving the use of HCN) for lengthening the carbon chains of sugars to convert pentoses into hexoses, the latter into heptoses, etc., thereby synthesizing sugars with as many as nine carbon atoms. Starting with formaldehyde (CH_2O) or glycerol, Fischer also synthesized the lower saccharides. He prepared a continuous series of "carbon hydrates" containing from one to nine carbons. In referring to the chain-lengthening process, Fischer pointed out that, with time, trouble and money to spare, it would be possible to "climb a further few rungs up this ladder."

Fischer polymerized monosaccharides to disaccharides an dextrin-type substances. He recognized the synthesis of starch and cellulose would be difficult. He was confident, however, that such syntheses are possible.

Amino Acids, Polypeptides, and Proteins. At the beginning of Fischer's research on amino acids and their polymers, thirteen amino acids had been obtained as hydrolysis products of proteins. Fischer discovered additional amino acids, synthesized several of them, and resolved the d and l forms by fractional crystallization of slats prepared from the benzoyl or formyl derivatives and the optically-active bases strychnine or brucine.

Fischer used Curtius' method to separate mixtures of amino acids from protein hydrolyzates by fractionally distilling their esters. he discovered valine, proline, and hydroxyproline in this manner.

By 1907 Fischer had prepared polypeptide (leucyl-triglycyl-leucyl-triglycyl-leucyl-octaglycylglycine) consisting of eighteen amino acid units and having a molecular weight of 1213. He eventually prepared a polypeptide containing nineteen amino acid units. Fischer suggested the peptide or amino linkage - CONH - was repeated in the polypeptide molecule.

By 1905, Fischer had investigated twenty-nine polypeptides and examined their behavior with various enzymes. He characterized proteins by the number, nature, and arrangement of the component amino acids. In 1916, Fischer reviewed his work on some one hundred polypeptides and cautioned that these represented only a small fraction of the possible combinations that might be found in natural proteins.

Enzymes. Fischer stated the action of enzymes depends to a large extent on the geometrical structure of the molecule to be attacked and that the two must match like lock and key. Maltase hydrolyzed alpha-methylglucoside but not beta-methylglucoside; emulsin hydrolyzed beta-methylglucoside but not alpha-methylglucoside. The synthetic sugars with three and nine carbons were converted into alcohol and carbon dioxide as readily as the six-carbon glucose.

Depsides and Tannins. In forming depsides (polymers of phenol carboxylic acids) and tannin (sugar esters of phenol carboxylic acids), Fischer prepared the alkyl carbonates of the phenol carboxylic acids. Treatment of these products with phosphorus pentachloride gave the corresponding acid chlorides; the latter were useful in preparing depsides and tannins. Using such reagents,

Fischer synthesized licanoric acid, a depside found in lichen substances.

A tannin-like material was prepared by treating tricarbomethoxygalloyl chloride with finely divided glucose in quinoline and chloroform. The resulting pentatricarbomethoxygalloyl glucose was hydrolyzed to pentagalloyl glucose, which was similar to tannin.

During the investigation of tannin-like substances, Fischer prepared a hepta (tribenzoyl galloyl)-p-iodophenyl-maltosazone, having a molecular weight greater than 4000.

Hydrazine Derivatives. Fischer prepared many organic derivatives of hydrazine, including phenylhydrazine, and explored their reaction. The reaction of hydrazines with carbon bisulfide gave dyes. Oxidation led to tetrazenes, which have four nitrogen atoms. Arylhydrazines reacted with ketones to give indoles (Fischer indole synthesis, 1886). Fischer discovered in 1884 that Phenylhydrazine is a valuable reagent)forming crystalline hydrazones an osazones) for investigating sugars.

Purines. Fischer investigated uric acid and related purines during the period 1881-1914; in 1914 he achieved the first synthesis of a nucleotide. He investigated the entire series of purines, established their structure, and synthesized the parent substance (purine) and about 130 derivatives.

Industrial Products. Fischer's thesis leading to the doctorate at Strasbourg in 1874, was concerned with phthalein dyes, particularly fluorescein and orcin-phthalein. At Munich, Emil and cousin Otto Fischer investigated rosaniline, the dye prepared by A. von Hofman in 1862 by the oxidation of aniline and toluidine. They converted one of the rosaniline dyes to triphenylmethane, and proved in 1878 that the various rosaniline dyes are triamine derivatives of triphenylmethane (4).

Fischer's purine research also was of interest to the German chemical industry. His laboratory methods became the basis for the industrial production of cafeine, theophylline, and theobromine. In 1903, he synthesized 5,5-diethylbarbituric acid. Under various trade names - Barbital, Veronal, and Dorminal - this compound proved to be a valuable hypnotic. Another commercially valuable purine was phenylethylbarbituric acid prepared by Fischer in 1912 and known as Luminal or Phenobarbital (4).

Fischer's phenylhydrazine was an integral link in the synthesis of antipyrine, a successful febrifuge, carried out by his own student and friend, Ludwig Knorr (1859-1921) (7).

Glycerol Esters. Toward the end of his career, Fischer turned his attention to the fats. Although these researches were not completed, his investigation of the simple glycerol esters produced an important discovery, namely, the acyl groups in esters of aliphatic polyhydroxy compounds may wander. The titles of some of his articles in this field are "Migration of the acyl in glycerides," "Interchangeability of esters and alcohol groups in the presence of catalyzers," and "New Synthesis of Monoglycerides."

FISCHER'S SON

Emil Fischer's oldest son, Hermann Otto Laurenz (1886-1960), also had a distinguished career in chemistry. Educated at Cambridge, Berlin, and Jena, Hermann went to Berlin in 1912 to begin research under his father's guidance. He served as professor of research at the

Chemical Institute of Berlin University, the University of Basel (1932), The Banting Institute of the University of Toronto (1937), and the new Biochemistry Department at the University of California in Berkeley (1948). Hermann O.L. Fischer received the Sugar Research Award of the American Chemical Society in 1949, and the Adolf von Baeyer Memorial Gold Medal from the Society of German Chemists in 1955. He was named Professor Emeritus by the University of California in 1956.

EMIL'S LAST YEARS

Emil Fischer's successes, satisfactions, and triumphs were accompanied by disappointments and tragedies. The gastric catarrh that forced him at the age of eighteen to delay the beginning of his higher education struck again and caused him to take leave of absence for one year when he was at Erlanger. His attractive wife, Agnes Gerlach Fischer, died after seven years of a happy marriage. Fischer suffered from mercury and phenylhydrazine poisoning and was in poor health during the latter part of his life. He lost two of his sons and various friends in World War I. Germany's defeat in the war was a bitter disappointment. After the close of the war, research activity was resumed. But cancer, that dread malignancy he had himself once tried to vanquish by chemotherapy, attacked him. When the diagnosis was confirmed, he put his house in order and committed suicide on July 15, 1919; his grave is in Wannsee near Berlin.

REFERENCES

1. Alyea, N.H., in Collier's Encyclopedia, Vol. 9, E.Friedman, Editor, P.F. Collier, Inc., New York, 1981.

2. Debus, A.G., World's Who Who in Science, Marquis-Who's Who Inc., Chicago, 1968.

3. Encyclopedia Britannica, Vol. 9, William Benton, Publisher, Chicago, 1966.

4. Farber, Eduard, in Dictionary of Scientific Biography, Vol. 5, C.C. Gillispie, Editor, Charles Schribners Sons, New York, 1972.

5. Forster, M., "Fischer Memorial Lecture," <u>J. Chem. Soc.</u>, <u>117</u>, 1157-1201 (1920).

6. Reingold, N., and Reingold, I.H., "Science in America: A Documentary History, 1900-1939, University of Chicago Press, Chicago, 1981.

7. Harrow, Benjamin, Eminent Chemists of Our Times, 2nd Ed., D. Van Nostrand Co., Inc., New York, 1927.

8. Hudson, C.L., <u>J. Chem. Ed.</u>, <u>18</u>, 353-357 (1941).

9. Ihde, A.J., The Development of Modern Chemistry, Harper and Row, New York, 1964.

10. Knight, David, in Encyclopedia Americana, International Edition, Vol. 11, Grolier, Inc., Danbury, CT, 1982.

11. Mark, H.F., <u>Chem. & Eng. News</u>, April 6, 1976, p. 176-189.

12. Mark, H.F., <u>Science, 146</u>, Nov. 20, 1964, p. 1023-4.

13. Mark, H.F., <u>Impact of Science on Society</u>, <u>18</u>, 27-33 (1968).

14. Mark, H.F., History of Polymer Science and Technology, R.B. Seymour, Editor, Marcel Dekker, Inc., New York, 1982.

15. McConnell, V.F., in Academic American Encyclopedia, Grolier, Inc., Danbury, CT, 1982.

16. Moore, F.J., A History of Chemistry, 3rd Ed., McGraw-Hill Book Co., New York, 1939.

17. Darmstaedter, L., and Oesper, R.E., J. Chem. Ed., 5, 37 (1928).

18. Theel, H., Nobel Lectures, 1901-21, Elsevier Publishing Co., New York, 1966.

19. Farrar, W.F., in A Biographical Dictionary of Scientists, T.I. Williams, Editor, John Wiley & Sons, New York, 1982.

20. Helferich, B., in Great Chemists, Eduard Farber, Editor, Interscience Publishers, New York, 1961.

21. Fischer, E., Nobel Lectures, 1901-21, Elsevier Publishing Co., New York, 1966.

22. Stahl, G.A., in Polymer Science Overviews, A Tribute to Herman F. Mark, G.A. Stahl, Editor, American Chemical Society, Washington, D.C., 1976.

23. Skolnik, H., and Reese, K.M., A Century of Chemistry, American Chemical Society, Washington, D.C., 1976.

24. Fischer, Emil, J. Am. Chem. Soc., 36, 1170 (1914).

25. Hosch, K., Ber., 54, 480, (1921).

CHAPTER 7

THE DEVELOPMENT OF THERMOSETS
BY LEE BAEKELAND
AND OTHER EARLY 20TH CENTURY CHEMISTS

ABSTRACT

The invention of Velox photographic paper, for which he received $1 million dollars, would have established Leo Baekeland as a great inventor. However, he enhanced his reputation as an inventor and chemist by financing his own research which led to the development of Bakelite, the first synthetic plastic.

This Belgian born, Ghent educated, scientist recognized that the formation of insoluble "goos, gunks and messes" could be controlled by an understanding of fundamentals. By controlling the ratio of trifunctional phenol and difunctional formaldehyde, he was able to produce a linear thermoplastic prepolymer which could be converted to an infusible crosslinked polymer.

In addition to producing a plastic, which is now manufactured in the US as an annual rate of over 1 million tons, he established the techniques which were used to produce urea, melamine, epoxy and polyester plastics.

Leo Baekeland also demonstrated his professionalism by serving as President of the American Chemical Society.

EARLY DEVELOPMENTS IN THERMOSETTING PLASTICS

Some scientists credit John Wesley Hyatt with the invention of synthetic plastics. However, ebonite, which is

R. B. Seymour (ed.), Pioneers in Polymer Science, 81–92.

a thermoset polymer with a high crosslink density, was invented by Nelson Goodyear in 1851, several years before Hyatt added camphor to cellulose nitrate in order to mold this tough thermoplastic.

Of course, the first manmade crosslinked polymer was Charles Goodyear's vulcanized rubber with low crosslink density.   Since there were only a few sulfur crosslinks between the polyisoprene chains, Charles Goodyear's product, which he called vulcanite, was elastic but had a higher modulus than the original linear Hevea braziliensis. However, Nelson Goodyear's product, called ebonite, produced by the addition of larger amounts of sulfur (25-30%) to natural rubber, was a nonelastic intractable solid.

While the exact mechanism for crosslinking with sulfur is not known, the major crosslinks are believed to be on the allylic carbon atoms.   Neither Charles nor Nelson Goodyear were scientists and they had little knowledge of crosslinking or of macromolecular structure.   Nevertheless, they made monumental discoveries as witnessed by the fact that sulfur continues to be the major curing agent for both natural and synthetic rubber.

It should be noted that Charles Goodyear died penniless and was not absolved of his crime of having wasted money in attempts to protect his vulcanization patent, until 1982.   Goodyear was involved in 60 infringement suits having been awarded the vulcanization patent in 1839. His last trial lawyer, Daniel Webster, won his patent infringement suit in 1852.

Goodyear was even jailed in debtor's prison in France where he was awarded the Legion of Honor Medal by Napoleon III.   The name vulcanization was coined by Goodyear after the Roman god of fire, Vulcan, and has been used subsequently, by many other scientists.

Possibly, more important, the world's largest rubber and tire company was named after the impoverished inventor of the vulcanization process.

While there were few other noteworthy scientific developments during the life of Charles Goodyear, other inventors used the vulcanization process for the production of useful end products, such as pneumatic tires. A Scottish chemist, named Charles MacIntosh, produced waterproof coats by placing a solution of natural rubber in naphtha, between two pieces of fabric. The garment still bears his name.

Another Scotsman, named John B. Dunlop, replaced solid rubber bicycle tires by air-filled rubber tires in 1887. Bicycle and automobile riders would not have been able to ride comfortably without the reinvention of the pneumatic tire by this Scottish veterinarian.

Berzelius obtained polyester resins by the condensation of polybasic acids and polyhydric alcohols in 1847 and J.J. Brenninlar investigated similar polyesters in 1856. However, these polyester resins were not commercialized until 1901 when Watson Smith obtained Glyptal resins by the reaction of glycerol and phthalic anhydride.

The previous investigation on phenolic resins by von Bayer, Kleeberg, Smith, Luft, Faolle, DeLaire, Lederer, Storey, Manasse, Speyer, Grognot, Helm, Knoll and Lebach were well known to Leo H. Baekeland when, at the age of 35, he decided to make a phenol formaldehyde resin as a replacement for shellac. In contrast to the Goodyears and other early pioneers in polymer technology, who were not trained as scientists, Baekeland was a Ph.D. chemist with a successful track record of scientific inventions.

This practical polymer chemist was born in Ghent, Belgium on November 14, 1863. As was the European custom, Leo received all of this education in his home town. He graduated, with honor, from the Municipal Technical School of Ghent in 1880 and received a scholarship at the University of Ghent. He was awarded the degree of Doctor of Natural Science Maxima cum laude from that university in 1884 at the age of 21.

After serving three years as an associate professor of chemistry and physics at the Government School of Science, he was awarded a three year travelling fellowship in competition with the graduates of three other Belgian universities. He served for two years as a professor of chemistry at the University of Bruges before taking his first trip to the U.S. in 1889. Leo married Celine Swarts, the daughter of Professor Theodore Swarts, who succeeded August Kekule at Ghent.

While visiting Columbia University, Baekeland was encouraged by Professor Chandler to remain in the U.S. Accordingly, Leo and Celine remained and eventually became citizens of the U.S. However, Leo retained his friendship with his old Belgian colleagues and while he was bilingual, he never lost his Flemish accent. The Baekelands were parents of two children.

Dr. Baekeland's first invention was in photography and the first of his 55 American patents was granted in this field in 1887. He established the Nepera Chemical Co. to manufacture a new "gas light" photographic paper in 1891. This paper, which he called "Velox" was the first photographic paper that could be developed under artificial light.

George Eastman, who patented various types of photographic film, including cellulose nitrate (Celluloid) as

well as the "Kodak" camera, purchased the Velox patent
from Baekeland for $1 million. Leo had anticipated asking
$50 thousand and would have settles for $25 thousand but
George Eastman made his offer before Baekeland could
state his price.

At the age of 36, Dr. Baekeland used part of the
"Velox" money to build a small laboratory adjacent to his
home ("Snug Rock") at Yonkers, NY. His second major
invention was an electrolytic cell for the production of
chlorine and caustic soda. After successful pilot plant
runs, this cell was used to establish the Hooker Chemical
Co. at Niagara Falls, NY. It is of interest to note that his
research associate, C.P. Townsend, who was coinventor of
the "Townsend Cell" served as Baekeland's patent attorney
for many years.

PHENOLIC RESINS

While working with another associate, Nathanial
Thurlow, Baekeland developed techniques for controlling
the condensation of phenol and formaldehyde. He wrote
little about the science of polymers, yet, it is evident, from
his first publication in Industrial and Engineering
Chemistry in 1909, that he had a better understanding of
functionality than his illustrious predecessors.

His "heat and pressure" patent demonstrated that he
recognized the need to maintain pressure on the resin
while converting it from an A and B stage to a C stage
infusible plastic. The A stage resin was produced by the
condensation of phenol and formaldehyde in the presence
of an alkali.

Baekeland used the term resole to describe resins
made with alkaline catalysts. Those made with acid
catalysts were called novolacs. However, the A stage

novolac was obtained by the reaction of phenol with less than a mole equivalent of formaldehyde. The fusible linear novolac resin was converted to a C stage in the mold by the addition of hexamethylenetetramine. Baekeland first called his resin, oxy-benzyl methylene-glycol anhydride but registered the trade mark of "Bakelite".

Leo Baekeland demonstrated that he was a professional chemist by discussing his inventions before the New York Section of the American Chemical Society in 1909. He was the recipient of the first Chandler award in 1914 and served as national president of the ACS in 1924.

The first commercial Bakelite resins were made in a small kettle in Baekeland's garage using steam both for heat and for driving the agitator. The General Bakelite company plant was started in a leased building in Perth Amboy, NY in 1911. This company became the Bakelite Corp. in 1921.

It is of interest to note that George Eastman used Bakelite for the end panels of his Kodak camera in 1914 and that the Hyatt Burroughs Billiard Ball Co. replaced Celluloid with Bakelite for its billiard balls in 1912.

The Bakelite billiard balls were cast in glass lamp bulbs and presumably this attracted the attention of the inventor of the electric light. Thomas Edison who made cylindrical phonograph records was seeking a new plastic for making flat disc records.

Edison's chief chemist, J.W. Aylesworth produced phenolic resins which were not suitable for molding records. Nevertheless, Aylesworth, along with Dyer and Kirk Brown formed the Condensite Company of America for the production of phenolic molding resins.

Dr. Baekeland granted a license to Condensite but would not grant one to the Redmanol Chemical Products Co. The latter was formed by Adolph Karpen for the production of phenolic resins, developed by Dr. Lawrence V. Redman and his associates, Frank P. Brock and Archie J. Weith.

Baekeland sued Redman and obtained patent infringement judgement which prevented Redmanol from making resins. However, Karpen purchased stock in Condensite Corp. which had been granted a license by Baekeland. The controversy was resolved by merging the three competitors into the Bakelite Corp. in 1922.

When Union Carbide and Carbon Corp. purchased the Bakelite Corp. in 1939, Kirk Brown's sons, Sanford, Allan and Gordon as well as C.P. Townsend joined the parent company. Leo Baekeland continued his activities as a scientist and as an executive. he served as an Honorary Professor of Chemical Engineering at Columbia University in 1917 and served on the U.S. Naval Consulting Board.

He made wine from his grapes at Snug Rock and brewed his own beer; both before and during the Prohibition Era. By application of his scientific knowledge of small molecules, he was able to produce wine with 18 percent alcohol and beer with 12 percent alcohol. The Memorial Resolution to Leo Hendrik Baekeland from Columbia University ends with the statement: "No man knew better how to live usefully, triumphantly and beautifully than did Leo Baekeland."

While wood flour-filled-phenolic resins were used for molding electrical components, and phenolic resin-impregnated paper was used for laminates (Formica, Micarta), attempts to use phenolic resins in oleoresinous varnishes were unsuccessful. This problem was solved by

K. Albert who produced Albertol by heating phenolic resin with rosin and esterifying the rosin with glycerol. "Four hour enamels" were produced in 1926 by use of a combination of Albertol and tung oil.

Oil soluble linear phenolic resins were produced in 1928 by the reaction of formaldehyde with ortho or para-substituted phenols.   E.E. Novotny patented phenolic resins, produced by the condensation of phenol with furfural.

AMINO RESINS

The successful use of resins produced by the condensation of formaldehyde with phenol catalyzed the development of other condensation products of formaldehyde with urea and melamine. The urea resins were described by Tollens in 1884 and were patented by John in 1918.

Contributions to the urea resins technology were made by Goldschmidt and Neuss in 1921 and by Pollak and Ripper in 1923.   Molding resins, based on condensation products of urea and thiourea with formaldehyde (Beetle) were produced in England in 1926.   Toledo Scales introduced a urea-formaldehyde molding powder (Plaskon) in 1928.   Paper impregnated with urea resin was used as the outer surface layer of Formica decorative laminates in 1931.

Resins produced by the condensation of formaldehyde with melamine were introduced by Ciba in 1933 and patented by Henkel in 1936.   Comparable resins were produced by Palmer Griffith of American Cyanamide Corp. in 1933.   This firm produced alpha cellulose-filled melamine molding compounds in 1937 under the trade names of Cymel and Resimene. The annual production of amino resins in the U.S. is 1.5 billion pounds.

POLYESTER RESINS

The growth of Glyptal resins was hampered by high cost and insufficient supply of phthalic anhydride. This problem was solved in 1936 by the development of the Gibbs' catalytic process for the oxidation of naphthalene. R.H. Kienle of General Electric Co. modified Glyptal resins with drying oils and coined the named "alkyd". Kienle's patent for alkyd resins obtained in 1933 was declared invalid in 1935 but the acronym alkyd continues to be used to describe this major coating resin.

Glycyl maleates were described by D. Vorlander in 1894 and synthesized by W.H. Carothers in 1929. Carlton Ellis was granted a patent for the polymerization of a solution of glycol maleate in vinyl acetate in 1937. Because of its volatility, the vinyl acetate was replaced later by styrene and these systems were used to impregnate fiberglass for the production of polyester composites. The annual production of unsaturated polyester resins in the U.S. is 1.4 billion pounds.

EPOXY RESINS

Patents were granted to J. MacIntosh and E.Y. Walford for diepoxide compositions in 1920. This reaction with phenol was extended to diols, such as bis-phenol A by Schlack in 1935, Castan in 1939 and Greenlee in 1940. Of course, many improvement patents have been granted on these epoxy resins since that time. The annual production of epoxy resins in the U.S. is 400 million pounds.

THERMOSET CHEMISTRY

It is of interest to note that all of the major thermosetting plastics were developed before 1940 when the science of macromolecules was in its infancy. The early developments, such as the production of vulcanized rubber with low and high crosslink density was by nonchemists who were unaware of polymer science. However, Baekeland, Redman, Albert, Kienle and Greenlee and many of their contemporaries were competent organic chemists who were aware of the developments in polymer science both through their own efforts and those of Staudinger and Mark.

Baekeland and most of those who pioneered the development of thermosets were practical chemists whose livelihood was dependent on the commercial production of these polymers. Since these products were crosslinked, they did not require the high purity that was essential for the production of linear condensation polymers.

All of these pioneers were actively engaged in the production of polymers of almost infinite molecular weight when Staudinger described macromolecules at the Dusseldorf meeting in 1925. Staudinger stated that he had established the existence of molecules that are a thousand times larger than those that his antagonists were studying in their laboratories. While antagonists, such as Karrer, Pringsheim, Bergman and Waldschmidt-Leitz were unreceptive to Staudinger's concept of macromolecules, Baekeland and his contemporaries were producing these giant molecules commercially.

While Carothers worked primarily with linear macromolecules, he confirmed Staudinger's concepts in the early 1930's and was primarily responsible for the development of the synthetic fiber industry at a time when thermosetting plastics were essentially the only synthetic polymers available.

BAEKELAND - OUTSIDE THE LABORATORY

Leo Baekeland outlived Wallace Carothers, both in the length of time and in his manner of living. Leo was a pioneer motorist in 1899, an automobile tourist in Europe in 1907 and a yachtsman starting in 1899. In addition to writing technical articles on Bakelite, he also published a book "A Family Motor Tour through Europe" in 1908. This entrepreneur - chemist and bonlivant died at Beacon, NY, February 23, 1944.

REFERENCES

1.   Biographical Memoirs, Vol. 24 (L.H. Baekeland), p., 281, National Academy of Sciences, Columbia University Press, New York, 1960.

2.   Baekeland, L.H., U.S. Pats. 939,966 (1909), 942,852 (1909).

3.   Baekeland, L.H., Ind. Eng. Chem., 1, 3, 8, 149 (1909).

4.   Baekeland, L.H., Ind. Eng. Chem., 3, 932 (1912).

5.   Baekeland, L.H., Science, 40, 179 (1914).

6.   Baekeland, L.H., Ind. Eng. Chem., 16, 10 (1924).

7.   Baekeland, L.H., and Bener, H.L., Ind. Eng. Chem., 17, 3 (1925).

8.   Baekeland, L.H., J. Chem. Ed., 9, 1000 (1932).

9.   Cohoe, W.P., <u>Chem. & Eng. News</u>, <u>23</u>, 228, 276 (1945).

10.   Kendall, J., <u>Chem. & Eng. News</u>, <u>67</u>, (1949).

11.   Overberger. C.G., "Proceedings of the International Francqui-Colloquium," Brussels-Ghent, Nov. 1981.

12.   Redman, L.V., <u>Ind. Eng. Chem.</u>, <u>20</u>, 1274 (1928).

13.   Schofield, M.S., <u>Chemist Druggist</u>, <u>180</u> (4370) 53 (1965).

14.   Seymour, R.B.,, "History of the Development and Growth of Thermosetting Polymers," Chapter 7 in "History of Polymer Science and Technology," R.B. Seymour Ed., Marcel Dekker, Inc., New York, 1982.

CHAPTER 8

HERMANN STAUDINGER
FATHER OF MODERN POLYMER SCIENCE

Herman F. Mark
Polytechnic Institute of New York
Brooklyn, NY 11202

ABSTRACT

Staudinger's importance to the development of Polymer Chemistry rests on a threefold activity which he maintained with never failing enthusiasm for more than 30 years: as a scientist, as a teacher, and as a preacher. Guided by true scientific curiosity for the unknown, Staudinger selected as the work of his life, in the early 1930's, a field which, at that time, was hardly considered to be a worthy goal for an organic chemist of his reputation - the study of the natural organic substances of high molecular weight. Until then, Staudinger had cultivated typical problems of classical chemistry with well-defined substances which could be characterized by such standard methods as melting and boiling point, freezing point depression, and boiling point elevation.

High ideals, creative imagination, and hard work were never more splendidly and more deservedly rewarded than in the case of the many whose memory I am recalling for you.

HERMANN STAUDINGER

First of all, I am anxious to express my sincerest gratitude to the organizers of this Symposium, Distinguished Professor Raymond B. Seymour of the University of Southern Mississippi and Dr. Allan G. Stahl

R. B. Seymour (ed.), Pioneers in Polymer Science, 93–109.
© 1989 by Kluwer Academic Publishers.

of EXXON Research & Engineering   for giving me the welcome opportunity to tell you the story of one of the great representatives of our Science, whom I know almost from the year when he started to devote himself to "Macromolecules" and for whom I had been asked to write an obituary in 1966, one year after his death.

CLASSICAL ORGANIC CHEMIST

Hermann Staudinger was born in Worms in the Rhine on March 23, 1881 as the son of Franz Staudinger and Auguste (nee Wenck) Staudinger; his father was a Neo-Kantian philosopher; his older brother, Hans, became a famous Social Scientist and was for many years Professor and Dean at the New School of Social Research in New York City.   After graduating from high school in 1899, Hermann wanted to study botany because he loved plants and flowers then and, in fact, for the rest of his life. His father suggested that for that purpose he should also take some chemistry courses. His son followed this advice and got stuck with organic chemistry which he studied in Darmstadt, Munich and Halle, where, in 1903, he received his Ph.D. degree under the Professor O. Vorlaender with a dissertation   on   the   "Malonic   Esters   of   Unsaturated Compounds."

Chemistry and particularly organic chemistry was at this time a highly developed, sophisticated and respected science; since the days of Liebig, Hoffmann and Kekule to the present, most distinguished representatives, like Bayer, Fischer and Wilstatter, had succeeded in synthesizing many hundred thousand new compounds which had never before existed on this earth, these discoveries led to miraculously beneficial applications in all sectors of human life   and   activities;   such   as   agriculture,   education, recreation, medicine and hygiene.   They also had an impact on such industries as textiles, paper, packaging and transportation. At the same time, the molecular structure of innumerable reactive natural substances was established including alkaloids, vitamins, hormones, natural dyes, vegetable and animal poisons, shellac and saccharides. The newly promoted doctor, who concentrated his energies on problems of this kind, became assistant to Johannes Thiele at the University of Strassburg and published several studies on the reduction of carboxylic acids to aldehydes and the dechlorination of several alpha chlorinated fluorine derivatives. In the course of these investigations, he prepared the first "Ketene" which contained one carbon-carbon double bond attached to a carbon-oxygen double bond. This cumulation of unsaturation produces a special type of high chemical reactivity which was of considerable interest at that time and encouraged Staudinger to continue work on this group of substances. In 1907, he was named Professor at the Technical University in Karlsruhe as a successor to Helmut Scholl. Prominent scientists such as Carl Engles in Oil Chemistry, Fritz Haber in Chemical Engineering and K. Gaede in Physics were member of the faculty at Karlsruhe. The young Professor joined this group and soon started to publish a most successful series of classical investigations, all of which included interesting concepts and required unusual experimental skill. The work on ketenes was continued, the various chemical reactions of oxalylchloride were started and several aliphatic azo compounds were prepared and characterized. In Karlsruhe, Staudinger also carried out the polymerization of formaldehyde to polyoxymethylene together with C.L. Lautenschlaeger; this synthesis and a new method for the preparation of butadiene and isoprene, brought him, for the first time, to the field of polymerization processes. But his attachment to classical organic chemistry was not yet at its end: in 1912 he published his first book: "Die Ketene" at the "Verlag Enke" in Stuttgart. At this time Richard Willstatter, the acknowledged World Leader in

Organic Chemistry was appointed Director of the newly founded Kaiser Wilhelm Institute for Chemistry in Berlin and his position as Professor at the Eidgenoessische Technisch Hochschule (ETH) in Zurich became vacant. It is an impressive testimony to Staudinger's ability that this 31 year old chemist was selected as a successor to such a great man in preference to many much older contenders, each of which had already shown considerable capability and promise.

The move from Karlsruhe to Zurich - in 1912 - would have interrupted his research efforts but he was allowed to take several of his associates to the new position. After having established himself with his wife and two young children in the famous Swiss Metropolis, he studiously continued his work on Ketenes, nitrenes, phosphines and aliphatic diazo compounds but added new lines such as natural ingredients having special physiological activities. With Leopold Ruzika, he clarified the structure of natural pyrethrines, which were used as insecticides and devised the synthesis of an aromatic compound which served as "synthetic" pepper. In cooperation with Thaddaeus Reichstein, he carried out difficult and time-consuming analyses of the natural coffee aroma and succeeded eventually in preparing a surrogate with astonishingly close character. Thus, before and during the First World War, Staudinger's activities as teacher and researcher followed essentially the mainstream of organic chemistry of those days and provided important contributions to its affluence. Before reaching his fortieth year, he was already one of the acknowledged leaders in his profession, thoroughly familiar with many alluring approaches for the synthesis, characterization and application of organic substances. He was already the author of a highly practical manual on analytical procedures and of another, fundamentally oriented volume on a class of new and unusually reactive substances. There was also the probability that - soon - his career would be crowned by a call to one of the few and most prestigious positions in Chemistry, Berlin or Munich, and that he would continue to contribute to classical organic chemistry which he had started so successfully twenty years earlier.

NEW CONCEPTS, NEW VISTAS

But Staudinger's pioneering spirit moved him away from the peaceful continuation of known concepts and methods into new and revolutionary ideas and approaches. While he was still busily engaged in publications on butadiene, isoprene and dicyclopentadiene (1913-1924), on pepper taste and chemical constitution (1916-1924) and on insecticides (1922-1926), Staudinger performed the break with the traditional organic chemistry of these days with a 12 page publication "Ueber Polymerisation" which appeared on March 13, 1920 in the "Berichte der Deutschen Chemischen Gesellschaft", Volume 53. This article does not contain any new experimental data but represents essentially a careful review (more than 60 references on 12 pages) of a certain number of reactions in the course of which larger molecules are formed through the combination of smaller ones. he proposed to use the expression "Polymerisation" only in such cased where the individual small units are combined to the larger ones by "normal covalent bonds." At the end of rather lengthy discussion, there finally emerged the three formulas which made this paper famous and by their simplicity and logical appeal gave a prosperous and decisive start for the further development of the "Hochmolekulare Polymeristionsprodukte" high molecular weight polymers. The selection of the three representative is ingenious:

1.   Paraformaldehyde - now known as
      polyformaldehyde

2.   Metastyrol - now known as polystyrene

3.   Rubber - i.e. polyisoprene

The polymerization product of formaldehyde which was studiously investigated at that time as a synthetic resin, the formation of which was based on the opening of a carbon-oxygen double bond to give a chain of C and O atoms without any substituents; the material was hard, opaque (crystalline) high softening and difficulty soluble; another synthetic resin (polystyrene) was studied with great interest at that time.   This was produced by repetitive opening of a carbon-carbon double bond and represented a chain carrying a heavy substituent - the phenyl group; it was a hard material, brilliantly transparent (amorphous), relatively low softening and readily soluble.

The third was a natural substance of great actuality and interest, UI_2, rubber, for which many attempts for a successful synthesis had been made.   It was represented by a chain which contained carbon-carbon double bonds and carried a small substituent - the CH_3 group.   The material was soft, transparent, low softening and readily soluble in organic solvents.   Thus, the three proposed macromolecules covered a wide range of "resins" - synthetic and natural, hard and soft, crystalline and amorphous with simple carbon-carbon backbones and also with a carbon-oxygen skeleton.   For all of them the same structural principle was postulated; not by the reliance on any strict and cogent experimental evidence but by the power of its logical derivation and most of all by its disarming and convincing simplicity.

IMPACT

In 1922, I was asked to join the staff of the newly founded Kaiser Wilhelm Institut for Fiber Research in Berlin-Dahlem the Director of which was Professor R.O. Herzog, a prominent representative of organic chemical technology with emphasis on textiles. The Institute was supposed to provide basic information for the German Textile Fiber Industry, particularly for viscose and acetate rayon, wool and silk together with other natural products like starch, rubber and wood. Harzog's leading idea was to apply newly developed experimental methods - x-ray diffraction and infrared absorption - to the study of fibers and membranes in the solid state. In those days each of these important natural products was considered a large separate discipline and empirically investigated. There existed textbooks and handbooks on wood, cellulose, starch, wool, leather and rubber; each of these fields had its societies, conferences and symposia, there was clear distinction between protein chemistry, cellulose chemists, rubber chemists and the like. Each discipline was a world by itself, just like Jupiter and Saturn - before Copernicus. This all changed with the Staudinger concept of long covalently bonded chains as proposed in his 1920 article.

Cellulose and proteins were not mentioned by Staudinger in 1920 but fortunately in 1921 two papers appeared. The chain structure of cellulose was presented as one possible explanation of existing x-ray experiment in one article and in the other it was put forward on the basis of kinetic measurements that "the polysaccharide is built up of long chains." For us, working on natural products in Herzog's institute, everything became now clear and simple: all products of our studies, regardless of their specific chemical compositions, consisted of long chains. The exact chain length itself was, initially not of too deep concern - there certainly ought to be more than eight or ten, perhaps even more than one hundred repeating units. But the explanation for the immense variety of properties and behavior was obviously the microstructure of the chain and the type of substituents attached to it. They could be hydrophobic or hydrophilic, small or large, polar or nonpolar, acidic or basic and with these distinctions one was back again in the well known field of classical organic chemistry.

Staudinger had given us a new unifying principle which made things marvelously easy and clear and provided thoroughly accustomed directions for thinking ahead, planning new experiments and explaining success of failure! Half a century earlier the confusing variety and behavior of many thousands of synthetic dyes had been clarified and became explicable by the recognition of certain chromophobic systems. Once such a chromphore was prepared, almost any color of the spectrum was obtainable by simply adding auxochromic groups to it. Now, again, a unifying principle - long chain structure - provided the backbone for the behavior and the various substituents controlled the specific properties of individual substances. From the chemist's point of view, the difference between rubber and cellulose had become basically as small as that between a yellow and a blue dyestuff.

A unifying principle once proclaimed and accepted creates order where confusion reigned and understanding where uncertainty had prevailed: consider how the ideas of Lavoisier and Mendeleev influenced chemistry.

Our team at Herzog's Institute - R. Brill, I.R. Katz, M. Polanyi, K. Wissenberg and I - consisted of novices who readily and willingly accepted the new gospel and profited from it. But older organic chemists who had gained great experience through work, over many years, were reluctant to admit the existence of exorbitantly large molecules and the necessity for the explanation of the properties which are displayed by natural and synthetic polymers.

From 1920 to 1926 Staudinger added to the fundamental article of 1920 more than 25 publications; in 1922 he proposed the generic term "macromolecules" for all natural and synthetic substance of this class which together with "polymers" is now generally accepted. In the other papers, he reported additional evidence for the macromolecular character of rubber, polystyrene and polyindene.

In spite of these efforts prominent representatives of the organic chemistry of those days like M. Bergmann, K. Hess, P. Karrer, H. Pringsheim and R. Pummerer maintained strong opposition to Staudinger's views and took the position that the behavior of rubber, cellulose and proteins could adequately be explained by the assumption of small basic units which are held together not by normal covalent bonds but by strong intermolecular forces of aggregation or agglomeration.

Staudinger, faithful to true scientific tradition, attempted to resolve this difference in opinion not by argumentation and with the aid of new experiments carried out with methods of the classical organic chemistry. He selected for these tests; natural rubber, because C. Harries and R. Pummerer had postulated that rubber consisted of small cyclic isoprene units which were associated through "partial valencies" emerging from the double bond of the rubber. If the double bonds were removed by hydrogenation, these valences must disappear and a low molecular weight hydrocarbon - similar to a lube oil - should result. The tests, however, showed that the hydrogenated rubber was very similar to the original sample, somewhat less elastic and gave solutions of high viscosity. A partial reintroduction of the double bond led to a product which was similar to the hydrogenated material. Similar hydrogenation and dehydrogenation experiments with polystyrene, polyindene and polyanethole, gave additional information concerning the stability of these compounds. This again pointed to the existence of normal chemical bonds in the chains of these materials. These results together with observations made with a series of polyoxymethylenes were presented at three prominent international meetings; in Innsbruck in 1924, in Zurich in 1925 and in Duesseldorf in 1926. In all three cases, there was lively opposition on the part of the exponents of the aggregation concept and also by the leading classical crystallographers: viscosity was not a legitimate measure of true molecular size; it could just as well be the result of colloidal aggregation. The chemical reactions with polyhydrocarbons - rubber, polystyrene and others would not allow conclusions on such important natural polymers as cellulose, starch or proteins and, most significantly the small crystallographic cells of cellulose, silk and rubber could never accommodate macromolecules.

Staudinger remained unperturbed and decided to direct his work in the future especially to meet the arguments which were put up against him at the occasion of these conferences.

X-ray studies of polyoxymethylenes between 1926 and 1929 showed conclusively that the chain molecules of these polymers were passing through many crystallographic basic cells and clarified the relationship between molecular weight and x-ray data.

Systematic studies on cellulose and its derivatives were begun in 1929 and were continued with many coworkers, until the late 1950's. They provided many opportunities to prepare polymeric homologous series over a wide range of molecular weights and to study the relationships both on chemical composition and on such properties as solubility and crystallinity.

Extensive viscosity measurements of solutions of cellulose acetate, cellulose nitrate and polystyrene were started in 1930 and followed up by ultracentrifuge studies in 1934. Together they proved conclusively the existence of molecules with molecular weights of several hundred thousands and the possibility of using adequate viscosity measurements for the approximate determination of molecular weights.

During all these urgent activities, for the defense of the macromolecular hypothesis, Staudinger did not lose the view for farther reaching, longer range concepts. Already in 1926 he emphasized the importance of macromolecules for biochemistry and biology - after all our body consists essentially of proteins which are macromolecules and all life process involve reactions of and between macromolecules. Later he would come back to the same theme and , together with his wife Magda, create a new research discipline, the "Macromolecular Bioscience."

By the end of the 1920's Staudinger's efforts to prove the macromolecular concept by chemical reactions with polymers and by improved x-ray studies on cellulose, silk rubber and chitin, by systematic measurements with the ultracentrifuge and, perhaps most of all by the brilliant synthetic work of W.H. Carothers in the laboratories of the DuPont company. he succeeded in preparing polyesters and polyamides in the macromolecular range and determined their molecular weight by exact and reproducible end group analysis.

Thus, it was time for a comprehensive presentation of the entire field, which Staudinger published in 1932 in the

Verlag Springer in Berlin; it summarized on a broad front, the origin of the macromolecular concept and its emergence as a new branch of organic chemistry.

The next decade, until the beginning of World War II, was devoted to systematic studies of such synthetic polymers as polystyrene, polyisobutylene, polyvinyl and polyacrylic derivatives, a series of copolymers and of polyelectrolytes. At the same time the work on polyoxymethylenes and on polyethylene oxide was continued and expanded. It was an amazing broad front along which Staudinger, together with an ever increasing number of able and devoted associates, invaded and conquered the newly created field of macromolecular science.

Immediately after the war, which brought an unavoidable interruption, Staudinger continued his work as researcher and educator with indefatigable energy and concentration moving closer to the important merger of macromolecular science and biology. Three comprehensive volumes provide an impressive testimony for his deep and wide studies of those years:

1947 "Macromolecular Chemistry and Biology," Wepf and Company, Basel

1950 "Organic Colloid Chemistry," 2nd Ed., Vieweg, Braunschwig

1961 "Arbeitserinnerungen," Dr. S. Huthig, Heidelberg

Immediately after the war, Staudinger made another important step toward the development of the new branch of science by the foundation of "Die Makromolekulare Chemie," a journal which greatly assisted the spreading of knowledge and experience; it always has been and still is today one of the most important vehicles for scientific and technical publications under the Editorship of Dr. Werner Kern, one of Staudinger's oldest and most celebrated associates.

Traditionally, exceptional performances are accompanied by prestigious awards and honors; Staudinger received many throughout the years; a list of them is attached at the end of this article.

Another great and lasting achievement of Staudinger was that he succeeded in transmitting this spirit of adventurous thinking combined with sober experimental reliability to a large number of students, who sat at the master's feet and could not fail to become thoroughly persuaded that his macromolecular chemistry was a strong and big branch of the tree of their science, destined to grow just as fast and irresistibly as macromolecules themselves. With this idea, however, they carried with them into their positions in academic and industrial life the deep conviction that the growth of polymer chemistry would depend entirely on meticulous and careful experimentation with methods newly designed and developed for this field. A generation of leaders in high school, universities, and industrial laboratories emerged from Zurich and Freiburg, the "Highboroughs of High Polymers." With more than 200 able and devoted students and coworkers, Staudinger published more than 600 articles in the field of Macromolecular Science.

The third and, indeed, not the least important role which Staudinger had to assume as his work progressed, was the role of a preacher. Many of his colleagues in high academic positions remained for a long time skeptical and overcautious. They did not approve of the strong terms with which Staudinger elevated his own working field to a "new branch of organic chemistry," and they displayed mistrust in a number of his methods and results. It was this atmosphere of negative incredulity which Staudinger combated throughout the years without fatigue and impatience. Again and again - in conferences, symposia, and conventions - he stood up as chairman, lecturer, or discussion speaker. He developed his ideas, explained the coherent pattern of the new concepts, and defended his position against all attacks with all his ingenuity and enthusiasm.

We have both witnessed numerous, unforgettable occasions in the 1920's and 1930's when history of chemistry was made in the eloquent clashes between Staudinger and the representatives of the "aggregation theory of the small units." Holding firm to his main ideas and introducing modifications wherever the facts demanded them, Staudinger emerged from these battles as the grand old man of macromolecular chemistry, the Nobel prize winner, the honorary doctor of many famous institutions of higher learning as the fatherly friend of his pupils, and as the benevolent counselor of his colleagues.

High ideals, creative imagination and hard work were never more splendidly and more deservedly rewarded than in the case of Hermann Staudinger, the founder and apostle of "Macromolecular Science."

HONORS AND AWARDS

1928   Ehrenmitglied der Physikalischen Gesellschaft Zurich

1928   Ehrenmitglied des Akademischen Pharmazeuten-Vereins Zurich

1928   Ehrenmitglied des Physikalischen Vereins Frankfurt/Main

1929	Mitglied der Heidelberger Akademie
1930	Emil Fischer Gedenkmunze des Vereins Deutscher Chemiker
1931	LeBlanc Gedenmunze der Societe Chemique de France
1932	Mitglied der Kaiserlich Deutschen Akademie der Naturforscher zu Halle und Senator fur Baden
1932	Verleihung der Plakette der Deutschen Kautschukgesellschaft
1933	Cannizzaro Preis der Reale Accademia Nazionale dei Lincei, Rom
1933	Korresp. Mitglied der Gesellschaft der Wissenschaften zu Gottigen
1937	Auswartiges Mitglied der Konigl. Physiographischen Gesellschaft Lund
1937	Ehrenmitglied des Vereins Finnischer Chemiker, Helsinki
1941	Auswartiges Mitglied der Finnischen Akademie
1946	Korresp. Mitglied der Bayerischen Akademie der Wissenschaften, Munchen
1948	Ehrenmitglied des Verbandes der Chemischen Industrie Badens e.v.
1949	Mitglied des Vorstandsrates des Deutschen Museums Munchen

1950 — Verleihung der goldenen Ehrennadel des Vereins Finnischer Chemiker Helsinki

1951 — Verleihung des Dr. Ing. e.h. durch die Technische Hochschule Karlsruhe

1951 — Verleihung des Dr. rer.nat.h.c. durch die Universitat Mainz

1953 — Nobelpreis fur Chemie

1954 — Verleihung des Dr. [C] h.c. durch die Universitat Salamanca

1954 — Verleihung der Silbernen Medaille der Stadt Paris

1954 — Verleihung des Dr. chem. h.c. durch die Universitat Turin

1955 — Ernennung zum Ehrenburger der Stadt Freiburg/Brsg.

1955 — Verleihung des Dr. s.c. techn.h.c. durch die Eidgenossische Technische Hochschule Zurich

1956 — Ehrenmitglied der Gesellschaft Deutscher Chemiker

1957 — Ehrenmitglied der Gesellschaft fur makromolekulare Chemie Tokyo

1957 — Verleihung des Grossen Verdienstkreuzes mit Stern des Verdienstordens der Bundesrepublik Deutschland

1957   Erenenung zum Ehrenmitglied der Societe Chimique de France

1958   Ernennung zum "Grand Officier de l'Ordre de la Couronne" durch den Konig von Belgien

1958   Ehrenmitglied der Japanischen Chemischen Gesellschaft Tokyo (Chemical Society of Japan)

1959   Korresp. Mitglied des Institut de France, Academie des Sciences, Divisions des Membres libres et des applications de la science a l'industrie

1959   Verleihung des Dr.h.c. der Universitat Strasbourg

1962   Verleihung der Goldenen Ehren-Medaille des Vereins der Textilchemiker

1965   Verleihung des Grossen Verdienstkreuzes mit Stern und Schulterband des Verdienstordens der Bundesrepublik Deutschland

CHAPTER 9

J.C. PATRICK
FATHER OF AMERICAN SYNTHETIC ELASTOMERS.

ABSTRACT

While he was not a chemist, Dr. Joseph Cecil Patrick invented America's first synthetic rubber in the early 1920's when few scientists recognized the existence of polymers. Unlike his contemporaries, he didn't discard the brown insoluble gum that was produced when he attempted to hydrolyze ethylene dichloride with sodium polysulfide. He named this product Thiokol after the greek words for sulfur and gum. He solved commercial production problems by inventing the suspension polymerization process and solved the compounding problems by degrading the high molecular weight polymer to a low molecular weight liquid polymer. The latter is now the principal binder for solid propellants.

BIOGRAPHICAL DATA

Like many other American pioneers in polymer science, the inventor of America's first synthetic elastomer was born in the midwest. However, unlike the typical chemist, he was not inspired by some dedicated chemistry teacher and was actually a medical doctor and not a chemist.

Joseph Cecil Patrick was born on August 28, 1892 in Jefferson County, MO. In spite of the lack of adequate finances, he was determined to become a physician. He worked as a claim adjuster for Kansas City Electric Car System while attending classes at Hahnemann Medical School. However, he left, before completing his studies in

111

R. B. Seymour (ed.), Pioneers in Polymer Science, 111–118.
© 1989 by Kluwer Academic Publishers.

medical school and attempted to enroll in the US Air Corps in World War I.

Because of his poor health, he was unable to meet the physical requirements for combat service but because of a need for medics, he was accepted in the U.S. Army Medical Corps and was sent to a hospital in France as a medical sergeant.

Joe barely survived the influenza epidemic and was actually found in a coma surrounded by corpses awaiting shipment for enterment. His escape from mass burial may be accredited to a massive dose of quinine administered by a French army nurse.

In order to continue his medical education after World War I, he accepted a position as a public health inspector in Kansas City, MO. However, he obtained a higher paying position as an analytical chemist for Armour Packing Co. in Buenos Aires, Argentina within a year.

After returning to the U.S., he received his M.D. degree from Kansas City College of Medicine and Surgery in 1922. He married Leah Burns who gave birth to a son in 1923. Unfortunately, Mrs. Patrick died from tuberculosis in 1935. Joseph Patrick married Olive Hudson in 1937 and the second Mrs. Patrick became the mother of two daughters.

Dr. Patrick strayed from his medical practice and with a partner named Nathan M. Mnookin established Industrial Testing Laboratory, Inc. in Kansas City in the mid 1920's. In addition to analyses, "Doc"Patrick also developed a process for producing pectin by an enthanolic precipitation of apple residues from a vinegar factory. He received regular bonuses from the vinegar plant, which may have been as payment for the license to purchase ethanol for resale as beverage alcohol during the prohibition era.

THE INVENTION OF THIOKOL

The next major project for the Industrial Testing Laboratory was an attempt to improve the process of making ethylene glycol from ethylene via the hydrolysis of ethylene chlorohydrin. In his attempt to produce this "anti-freeze" by hydrolysis of ethylene dichloride with sodium polysulfide, "Doc" Patrick obtained a gummy gunk which he named "Thiokol" based on the greek word theion (sulfur) and kommi (gum).

Unlike many early twentieth century organic chemists, "Doc" Patrick did not discard the undesired gummy product. Instead, he registered the trade name Thiokol, made patent applications and wrote the equation for the formation of thiokol as follows:

$$Cl(CH_2)_2Cl + NaSxNa \rightarrow (-(CH_2)_2) Sx + 2NaCl$$

It is important to note that the discovery of the first American synthetic rubber was in the mid 1920's when few scientists besides Hermann Staudinger and Herman Mark recognized the existence of polymers. Patrick demonstrated his inventive ability further by using rubber compounding recipes to cure (vulcanize) and reinforce this vile smelling, solvent-resistant, brown gum.

"Doc" Patrick secured funds through R.E. Wilson from Standard Oil Company of Indiana to finance additional research on Thiokol. However, little progress was made in eliminating the odor or in developing useful large scale production techniques during the period (1927-1928) when Standard Oil was supplying funds.

Patrick worked on other research contracts, such as the development of "smoke flavored" salt, for the Western Salt Laboratory in 1928. He also analyzed some bootleg scotch whiskey for Mr. Bevis Longstreth who was president of the salt company. Later, Longstreth arranged for further funding of the Thiokol project by a New York firm of Case, Pomeroy, and Co.

Subsequently, Patrick, solved the production problem by using a novel suspension polymerization technique. The Thiokol suspension based on a magnesium hydroxide slurry was readily purified and coagulated by acidification in the same manner as the coagulation of natural rubber latex. Since the Prohibition Act had not been repealed, it was easy to obtain glass lined tanks from an abandoned Kansas City brewery and to start the large scale production of Thiokol in 1929.

EXODUS FROM KANSAS CITY

Unfortunately, "Doc" Patrick was a better scientist than a politician. He refused to pay graft to cronies of "Boss Tom" Pendergast. As a result, his Thiokol plant was closed down by the Pendergast machine.

Patrick and Longstreth demonstrated their versatility by relocating in a vacant rubber plant in Yardville, NJ in 1930. However, because of the objections to the obnoxious odors from the Thiokol plant by the area residents and labor problems, it was necessary to abandon this plant site. Nonetheless, two tons of Thiokol was produced in this plant and this was enough to establish it as a worthwhile commodity.

For example, as early as 1933, Thiokol was used as a cable sheathing by the Westchester (NY) Lighting Co. My first use of Thiokol was as a plasticizer for sulfur. The plasticized product had been invented by W.W. Dueker and C.R. Payne and used as a hot melt jointing material. It was my pleasure to meet "Doc" Patrick, Bevis Longstreth, and Sam Martin of the Thiokol Corp. in the 1930's when I was investigating plasticized sulfur for the Atlas Mineral Products Company where Dr. Payne was Director of Research.

LARGE SCALE PRODUCTION

Thiokol was produced on a relatively large scale by Dow Chemical Company in Midland, MI in the late 1930's and early 1940's but this production was transferred to Trenton, NJ after World War II. In an attempt to reduce the characteristic odor, "Doc" Patrick investigated many dichloro reactants.

The products obtained from 1,3-dichloropropane and from 1,4-dichlorobutane were more odoriferous than the traditional Thiokol and the product obtained from 1,4-dichloropentane was a low molecular weight cyclic product.

A good, less odoriferous, solvent resistant polymer, was obtained from dichlorodiethyl ether but the yield of this product was poor. The most successful reactant was dichlorodiethyl formal.

LIQUID THIOKOL

The major achievement of "Doc" Patrick was the controlled decomposition of Thiokol by sodium hydrosulfide and sodium sulfite. This low molecular weight Thiokol which was produced by Patrick and H.R. Ferguson could be readily compounded on a rubber mill and cured in a mold. The low molecular weight liquid polymer (LP-3) obtained by the decomposition technique became the binder for solid fuel rocket propellants in the 1940's.

"Doc" Patrick retired and moved to Florida in 1948.
However, he returned to Pennsylvania later and entered a
new career as a builder and investor.   In spite of his
brilliant research and his being "the father of American
synthetic rubber", "Doc" Patrick received little recognition
during his active years with Thiokol, Corp. However, the
use of Thiokol LP-3 as a binder for solid fuel rocket
propellants after World War II drew considerable attention
to this modest scientist-inventor.

He received the Goodyear and the Elliot Cresson
Medals in 1958.   He also was inducted into the
International Rubber Science Hall of Fame, posthumously,
in 1981. He died in 1965.

CONCLUSIONS

Thiokol Corporation, which is the exclusive producer of
polysulfide elastomers in the free world, is now a major
producer of solid rocket engines and has annual sales of
many millions of dollars. Thiokol (now Morton-Thiokol) is
the sole supplier of solid rocket boosters for the space
shuttle mission and is listed on the New York Stock
Exchange.

The story of Thiokol is unique. It was developed by a
medical doctor who attempted to hydrolyze ethylene
dichloride with sodium polysulfide. The importance of this
pioneer american synthetic rubber was not recognized by
most chemists and this inventor was plagued by lack of
funds and essentially run out of town (Kansas City) by
Boss Pendergast.

He solved production problems by inventing the
suspension polymerization process. In his attempts to solve
the odor and processing problems, he developed the first
liquid elastomer which is now an essential component of

the Aerospace Program. Thus, in one lifetime, this sickly
scientist, not only made many scientific discoveries,
covered by more than fifty U.S. patents, but built up a
multimillion dollar polymer business.

REFERENCES

W.W. Duecker, Chem Met Eng 41 583 (1931).

T.P. Sager, Ind Eng Chem 29 742 (1937).

S.M. Martin, A.E. Laurence, Ind Eng Chem 35 986 (1943).

E. Fettes, J.S. Jorczak, Ind Eng Chem 42 2217 (1950).

J.S. Jorczk, E.M. Fettes, Ind Eng Chem 43 324 (1951).

E.R. Bartozzi, Rubber Chem and Technol 41 (1) 114
(1968), Rubber World 139 84 (1958).

J.C. Patrick, Modern Plastics 14 96 (1936).

J.C. Patrick, Trans Faraday Soc 32 347 (1936).

J.C. Patrick, Rubber Chem and Technol 9 373 (1936).

J.C. Patrick, Modern Plastics 15 36 (1937).

J.C. Patrick, U.S. Patent 1,854,423 (Thiokol).

J.C. Patrick, N.M. Mnookin, U.S. Patent 1,890,231
(Thiokol).

J.C. Patrick, U.S. Patent 1,950,744 (Suspension Process).

J.C. Patrick, U.S. Patent 2,469,404 (Dithioglycol formal
Polymer).

J.C. Patrick, U.S. Patent 2,479,542 (Liquid Polymers).

CHAPTER 10

Waldo Lionsbury Semon
Pioneer in PVC

Polyvinyl chloride (PVC) has also been classified as a mature thermoplastic but its use also continues to increase. Over 9 billion pounds of PVC and related vinyl plastics were produced in the US in 1987. PVC is one of the few commercial plastics that contains more than 50 percent of nonhydrocarbon constituents. Vinyl chloride, like styrene, remained a laboratory curiosity for a almost a century before it was introduced as a commercial product.

Regnault who synthesized vinyl chloride in 1835 was considered by Liebig to be one of the most talented students at Ecole Polytechnic in Paris. Regnault's synthesis, which was used commercially a century later, was based on the dehydrochlorination of the "oil of the Dutch chemists" (Dieman, Trotswyck, Bondt and Laurverenburgh) which was first produced in 1795, by the addition of chlorine to "olefiant gas". Kolbe used the term vinyl radical in 1854 and the term vinyl chloride was used without controversy in the 1860's.

In 1860, Hofman described the "metamorphosis" of vinyl bromide to polyvinyl bromide and Baumann repeated this "conversion" with vinyl chloride in 1872. However, polyvinyl chloride (PVC) was not patented until 1912, when Klatte used sunlight to initiate the polymerization of vinyl chloride. Klatte produced the monomer by the mercuric chloride catalyzed addition of HCl to acetylene. He also suggested the use of camphor and triphenyl phosphate as plasticizers for PVC.

In 1926, Ostromislensky patented flexible film cast from a solution of PVC and a plasticizer, such as

119

R. B. Seymour (ed.), Pioneers in Polymer Science, 119–122.
© 1989 by Kluwer Academic Publishers.

chloronaphthalene in chlorobenzene. However, he did not include phenyl phthalate or diethyl phthalate which had been patented as plasticizers by Clarke and Wilkie, respectively, in 1920 and 1922.

The first patent on a moldable plasticized PVC was granted to Wolfe of B.F. Goodrich in 1932. Nevertheless, the key patent which was responsible for the commercialization of PVC was granted to Semon in the same year.

As a result of the success of this commercial plasticized PVC (Koroseal), other firms such as DuPont, Union Carbide and Goodyear attempted to develop competitive products. DuPont abandoned this investigation but Reid of Carbon and Carbide Co. patented useful copolymers of vinyl chloride and vinyl acetate which he called Vinylite and these are still used today.

One of my assignments at Goodyear was to produce a coating for silk which could be competitive with Waldo Semon's Koroseal. Hence, the more amorphous and more soluble copolymer of vinyl chloride and vinylidene chloride was produced and patented in 1937 and given the trade name of Pliovic.

Since there was a great deal of secrecy among chemists employed by the rubber companies in Akron, Waldo and I never discussed our PVC research. However, since the technical employees in the Akron area were unusually social, we did meet frequently at the social affairs sponsored by various technical societies. The first couple on the dance floor was usually Waldo and Marjorie Semon.

Waldo Lionsbury Semon was born in Demopolis, AL on September 20, 1989. Since his father was a civil engineer, who was transferred to the state of Washington. Waldo received all of his education in that state. The University of Washington awarded him a B.S., Ph.D. and honorary doctoral degrees in 1920, 1923 and 1946, respectively.

After being employed by Falkenburg County (WA) and Everett Gas Works, he returned to his alma mater as an instructor in chemistry but left in 1926 to accept a position with B.F. Goodrich where he remained until retirement in 1963 when he joined the faculty of Kent State University.

Dr. Semon was awarded 117 patents by the US Patent office and is the author of 40 reports in scientific journals. He is a member of the American Chemical Society, and the American Institute of Chemical Engineers. He is the recipient of the Goodyear medal (1946), the National Manufacturers Association Modern Pioneer medal (1940), the International Synthetic Rubber medal (1964), the Elliott Cresson medal (1964), the Morley medal (1968), the Midgely medal (1970) and was named a modern pioneer in polymer science in Polymer News (1982).

He and Marjorie were married in 1920. The Semons are parents of three daughters, Mary Marjerie and Constance. The Semons reside in Hudson, OH.

REFERENCES

Semon, W.L., Stahl, G.A., Chapt. 12 in "History of Polymer Science and Technology," Marcel Dekker, New York, NY, 1982.

Baumann, E., Ann., 163, 132 (1872).

Klatte, F., Ger. Pat. 281,877 (1913).

Regnault, H.V., Ann., 14, 22 (1835).

Hoffmann, A.W., Ann., 115, 271 (1860).

Ostromisklensky, I., U.S. Pat. 1,721,034 (1926).

Clarke, H.T., U.S. Pat. 1,398,939 (1920).

Wilkie, H.F., U.S. Pat. 1,449, 156 (1920).

Wolfe, J.E., U.S. Pat. 2,050,595 (1932).

Semon, W.L., U.S. Pat. 1,927,453 (1932).

Reid, E.W., U.S. Pat. 1,935,517 (1928).

Seymour, R.B., U.S. Pat. 2,348,154 (1944).

CHAPTER 11

RAYMOND F. BOYER
THERMOPLASTIC PIONEER

Neuman obtained styrene monomer by the distillation of storax (liquid amber) in the early 1800's. Subsequently, in 1839, Simon repeated this experiment and obtained an elemental analysis of the distillate which he called "styrol". Blyth and Hoffman heated styrol and produced a solid which they called metastyrol.

Blyth and Hoffman noted the high refractive power of the "metastyrol" and suggested its use for optical purposes. In 1866, Erlenmeyer showed that styrene was actually vinylbenzene and in 1869 Berthelot produced this monomer by the pyrolysis of ethylbenzene which he obtained by the condensation of ethylene and benzene.

In spite of its availability and the clarity of this brittle polymer, styrene monomer remained a laboratory curiosity for over a century. However, after Tschunker produced styrene-butadiene elastomeric copolymers (Buna-S), chemists at IG Farbenindustrie reinvestigated styrene homopolymers and several copolymers including styrene-co-acrylonitrile. Polystyrene was produced commercially in Germany in 1925.

Dow also became interested in the production of polystyrene and in addition to its cellulosic and vinylidene chloride projects, also supported a project on polystyrene. One of the polymer pioneers, who was responsible for the commercialization of polystyrene at Dow, was Raymond Boyer. Dow produced polystyrene commercially in the US in 1935. Both IG Farbenindustrie and Dow used the Berthelot synthesis for the production of the styrene monomer.

R. B. Seymour (ed.), Pioneers in Polymer Science, 123–126.
© *1989 by Kluwer Academic Publishers.*

It was my good fortune to meet Ray at the Gibson Island Conferences in the late 1930's and early 1940's. He and I were delegates chosen to represent Dow and Monsanto at these conferences which were named the Neil Gordon Conferences after the meeting site was moved from Maryland to New England.

Unfortunately, Ray and I were unable to exchange much information since both of our companies were potential competitors in the production of polystyrene. Dow built its production facilities at Freeport, TX and Monsanto built its plant at Texas City, TX. Dow became the leading producer of polystyrene in the US and Monsanto was a close second until it sold its production facilities in 1987. Over 6 billion pounds of polystyrenes produced annually in the US.

There were many different executives who championed polystyrene production intermittently at Monsanto during its 40 years of styrene and polystyrene production. However, Ray Boyer championed the production of polystyrene at Dow consistently for a period of over 40 years. He is recognized not only as a pioneer but also as the world's most knowledgeable polystyrene scientist. He was one of the few polymer scientists to be elected to the National Academy of Engineering (1978).

Ray was born in Denver, CO, on February 6, 1910, and was awarded B.S. and M.S. degrees by Case-Western Reserve University in 1933 and 1935, respectively. He was awarded an honorary D.S. degree by his alma mater in 1955. He served as a visiting professor at Case-Western in 1974 and continues to serve as an adjunct professor. After retiring from his 40 year tenure as a research scientist and the first research fellow at Dow in 1975, he joined the Midland Macromolecular Institute as an affiliate scientist. He has been granted 20 patents by the U.S. Patent Office and has published more than 120 scientific reports in scientific journals.

In addition to presenting seminars to 20 different universities in the U.S., he has presented seminars or lectures in England, Germany, Switzerland, Poland, USSR, Canada and Spain. He is a member of the American Chemical Society and has served as chair of three of its symposia at national meetings. He has also served as chairman of the advisory committee for the polymer division of the National Bureau of Standards and the Gordon Research Conferences as well as the American Physics Society.

Ray was named a Pioneer in Polymer Science by Polymer News and is the recipient of the International award of the Society of Plastics Engineers (1968), The Borden award of the American Chemical Society (1970) and the Swinburne award of the Plastics Institute of UK (1977).

There have been many other pioneers in polystyrene but no other has served this industry over as long a period and as well as Ray Boyer.

REFERENCES

Bundy, R.H., Boyer, R.F., ed., "Styrene, Its Polymers, Copolymers and Derivatives," Reinhold, New York, NY, 1952.

Simon, E., Ann., 31, 265 (1839).

Blyth, J., Hoffman, A.W., Ann., 53, 289 (1845).

Berthelot, M., Ann. Chem. Phys. [4], 16, 156 (1869).

Erlenmeyer, E., <u>Ann.</u>, <u>137</u>, 353 (1866).

Brighton, C.A., Pritchard, G., Skinner, G.A., "Styrene Polymers: Technology and Environmental Aspects," Applied Science Publishers, London, 1979.

Boyer, R.F., Chapt. 19 in "History of Polymer Science and Technology," Marcel Dekker, New York, NY, 1982.

CHAPTER 12

WALLACE HUME CAROTHERS
INNOVATOR, MOTIVATOR, PIONEER

C.S. Marvel
Department of Chemistry
University of Arizona
Tucson, AZ 85721
and
Charles E. Carraher, Jr.
Department of Chemistry
Florida Atlantic University
Boca Raton, FL 33432

ABSTRACT

Wallace Hume Carothers is the proclaimed father of organic polymer science. Midwest bred,he packed more than a lifetime of accomplishments into 40 years of life. Carothers was a hard driving, intense, yet quiet and warm individual, reared in the midlands of America. Recognized by such "talent scouts" as Arthur Pardee, Roger Adams and Carl (Speed) Marvel, he moved to Harvard and then to du Pont where he established an empire that remains. Insights gleaned from first and second hand encounters will be related with an emphasis on both his emergence from the midlands to the establishing of the basic science of macromolecular synthesis.

PROLOGUE

History is measured by the change in the flow of grains of sand in the hour glass of eternity. Most of us are small contributing grains, but a few affect the lifestyle of many, being granite boulders in this seashore of humanity. Wallace Hume Carothers (1896-1937) is one

R. B. Seymour (ed.), Pioneers in Polymer Science, 127–143.
© 1989 by Kluwer Academic Publishers.

such boulder; his contributions contribute to the basic elements of the twentieth century - brushes for our hair and painting, socks for our feet, tires for our vehicles, and thread for our clothing.

This review is taken from the memories of many including Herman Mark, Julian Hill, Charles Estee, Paul Flory, Paul Morgan publications noted as references and research done by Fred Pease.

DEVELOPING YEARS

Wallace Hume Carothers was born, reared and educated in the midwest. His paternal forebears were of Scotch origin and settled in Pennsylvania in prerevolutionary days. They were farmers and artisans. His father, Ira Hume Carothers, was born in 1869 on a farm in Illinois. He taught a short time in a country school at the age of 19. Later he entered the field of commercial education as a teacher and later vice president of the Capital City Commercial College, Des Moines, Iowa.

His mother was Mary Evalna McMullin of Burlington, Iowa. Wallace was born in Burlington, Iowa on April 27, 1896, the oldest of four children. His sister Isobel (Mrs. Isobel Carothers Berolzheimer) was the Lu in the popular radio trio Clara, Lu and Em. His education began in the public school of Des Moines and in 1914 he graduated from North High School.

As a youth Wallace exhibited a zeal for work as well as play. He enjoyed tools and mechanical things, spending much time in experimenting. He was thorough in his classwork and did not begin tasks which he did not finish.

In the 1910's throughout the 30's there existed a dualism with respect to post-high school education. For some the steps to the "big times" were rapid with graduation from high school directly followed by entrance to college, etc. till the Ph.D. was granted. Then a choice between industry or a teaching position at one of a number of Ph.D. granting institutions. For others the road was slower and less sure. For some, work after each sojourn in school - graduation from high school, work, entrance to a smaller college such as Sterling, Friends of Tarkio, then to industry or enrollment to a masters' granting institution such as the University of South Dakota; in fact, the University of South Dakota acted as a transition school for many instructors with some deciding on a teaching career at a smaller institution and others (obtaining a doctorate if not already at the doctoral level) moving on to more research-oriented setting. It was Wallace's role in life to trudge this second, less sure route. In fact Wallace was 28 when he received his Ph.D., several years older than most of the "superstars" which he was to compete with. This may have contributed to his need to overachieve and in later years, his drive to both achieve the need to be accepted as a scientist of the first magnitude.

In the fall of 1914 he entered the Capital City Commercial College, graduating in July, 1915, specializing in the accountancy and secretarial curriculum. He then entered Tarkio College, a Presbyterian Christian - liberal arts college located in Tarkio, Missouri, in September, 1915 to begin a course in science, but simultaneously accepting a position as an assistant in the commercial department, to provide funds for his schooling. Though he had majored in chemistry since his arrival at Tarkio, he was made an assistant in english after two years. World War I was in full swing and the United States had entered into the battle. Wallace was rejected as a soldier due to a slight physical defect.

Dr. Arthur M. Pardee was called to the University of South Dakota to be chair of the Department of Chemistry. Tarkio College was in need of an instructor and Carothers, who had previously passed all of the chemistry courses offered, was appointed instructor of chemistry. It is noteworthy that during his senior year, there were four senior chemistry-major students in the class and all later completed work through the doctoral level. This bears testimony to the inspiration and leadership of Carothers. Pardee was a talent scout and served as such for many years at USD, selecting from often untried and unproved instructors encouraging them and then forwarding them to other institutions. His initial project was not untried, nor unproved but he had already showed mature judgement in both his teaching and scholarship, but was in fact Wallace Carothers.

Carothers left Tarkio College in 1920 with his B.S. degree and enrolled in the University of Illinois where he completed the requirements for the M.A. degree in the summer of 1921. He then joined Pardee for the 1921/22 school term, teaching analytical and physical chemistry. He carefully prepared his presentations, and while not a forceful lecturer, he motivated and taught his students with enough skill to convince Pardee that this was indeed a prized talent which should be further refined with additional schooling.

The initial literary contribution by Carothers was derived from work done in the same laboratory in which co-author Carraher, some 50 years later, began his adventures into organometallic polymers. In the fall, the area surrounding Vermillion was a hunters' paradise with pheasant and quail in great abundance. But Wallace had taken the job with the intent to secure funds for further education, and besides he was busy developing some independent research problems. The new chemistry building (now called the Pardee Laboratories) beckoned stronger than his newly found friends. The winter Dakota winds, blowing against the large windows of Pardee Laboratory,and the laboratories 12 foot high ceilings hiding any stray heat, encouraged researchers to have idle Bunsen burners in operation. He was especially interested in a 1916 paper of Irving Langmuir on valence electrons and sought to expand it to organic chemistry. Pursuing this idea he conducted the laboratory work necessary to produce his first contribution, "The Isosterism of Phenyl Isocyanate and Diazobenzene-Imide" which appeared in the Journal of the American Chemical Society, 45 1734-1738 (1923). His second paper, "The Double Bond" (J. Amer. Chem. Soc. 45 2226-2236 (1924), also involved valence electrons. This paper contained the first clear, definite application on a workable basis of the electronic theory to organic chemistry. This landmark paper is often overlooked because of his later contributions, but it is one of the fundamental papers in organic chemistry. Both of these publications were published before he had received his Ph.D. degree.

He returned to the University of Illinois in 1922 to complete his studies, under Dr. Roger Adams, receiving his doctorate in 1924. His research involved the catalytic reduction of aldehydes with platinum oxide, platinum black and the effect of poisons and other agents on this catalyst. Three papers resulted from these studies. While he majored in organic chemistry, he did well in supporting subjects. This was true throughout his schooling, from early grammar school on.

At graduation he was considered by the staff as one of the most brilliant students who had ever been awarded the doctorate. He filled a temporary vacancy as an instructor in organic chemistry for the 1924-25 school term.

In 1926 Harvard University, in need of an instructor in organic chemistry, hired Carothers. He taught courses in experimental organic chemistry, structural chemistry, and elementary organic chemistry.

DISCOVERY

History is forged in a real time-space cavity where events only occur when the conditions are right - the people prepared. The person was Wallace Hume Carothers - a complex man; shy, inquisitive, intellectually bright and quick; self-demanding and an over-achiever; pleasant but at times despondent; had the ability to extend reactions on known information; a motivator.

The view championed by Staudinger that polymers were, in fact, macromolecules was beginning to be accepted though the contest was to continue for two more decades. In fact, the work of Carothers and co-workers did much to place the concept of macromolecules on firm scientific "ground". Further, there were many known organic reactions which served as the starting place for Carothers ingenuity - extending reactions known to occur with the synthesis of small molecules, to the synthesis of macromolecules.

In 1927 the DuPont Company reached a decision to begin a program of fundamental research, "without any regard or reference to commercial objectives". This was a radical departure from the common practice of the chemical industry then which supported potentially profitable ventures, leaving the "ivy covered" hall of academia to provide the necessary fundamental research.

Carothers was selected to head the research in organic chemistry at the new Experimental station at Wilmington, Delaware. The decision to leave Harvard was a difficult one. The Experimental Station offered a chance to expand his research aspirations well beyond those possible at Harvard, though he enjoyed his teaching, the opportunity to do only basic research won out.

Dr. James B. Conant, professor of organic chemistry at the time Carothers, was an instructor (later President of Harvard and the first head of the Atomic Energy Commission), extolled the contributions in his short stay at Harvard.

"His resignation from the faculty to accept an important position in the research laboratory of the DuPont Company was Harvard's loss but chemistry's gain".

The object of Carothers work at DuPont, which he defined explicitly at the outset, was to prepare molecules of known structure through the use of established reactions of organic chemistry, "and to investigate how the properties of these substances depended on constitution". He had an ability of recognizing the significant points in a project or paper and an ability to utilize this information on other problems.

Initially he sought to extend simple emanation and esterification reactions to produce long chained products through the use of difunctional reactants. The first studies involved use of dihydroxy and dicarboxylic reactants to react through esterification reactions.

His first products were waxy brittle solids with short chain lengths, i.e., a low number of repeating units. The Carothers' group recognized that the reaction was an equilibrium reaction and, in order to obtain longer chains, water must be removed. This was cleverly done using as assembly called the molecular still which was essentially a hot plate with a vacuum connected to aid in the removal of water. (Recall the younger Carothers' interest in tools and mechanical things.)

The length of the polyester was determined by the material balance, number of molecules of diacid and diol employed as well as the ability to remove the water as it is being formed. The nature of the end group was determined by the balance of reactants employed. Thus, reactions where there were more molecules of the diol compared to the diacid had unreacted alcohol end groups. In fact, the chain length and number of repeating units were determined by Carothers and coworkers using well known chemical analysis techniques with the result that they determined that their products were low polymers with the number of repeating units in the area of 25-50. This was eventually increased with values greater than 100. Customary and later truly long chained polyesters were produced employing modifications of Carothers' initial approach where the number of repeating units exceeded 1,000.

Through end group analyses and other analyses, the polyesters were proved unequivocally to be linear products containing both reactants connected through ester groups. This provided the ammunition needed by Staudinger to convince many of the previously unconvinced skeptics that polymers were, in fact, macromolecules and not simply aggregates of small molecules held together by undefined forces.

An analogous study involved applying the already well known Wurtz reaction to dihalo compounds such as dodecamethylene dibromide. This compound, when treated with sodium in solution formed chains containing multiples of ten methylene (CH_2) units. The forces holding these units of ten together were the same as those holding the methylene units of the starting compound together, i.e. dodecamethylene dibromide. Fractions were isolated where the number of repeating units was 2 through 10.

This again established several concepts. First, polymers could be formed by employing already known organic reactions except employing reactants which had more than one reactive group per molecule. Second, the forces bringing together the individual units are the same as those which hold together the starting materials. These forces are called primary covalent bonds.

This work was described in a series of papers published between 1929 through 1931, including a landmark review article published in Chemical Reviews in 1931. In this article Carothers laid the foundation of much of organic polymer science. He defined terms as condensation and addition of polymer processes; discussed polycondensation reactions, polymerizations involving ring opening of cyclic reactants, addition reactions; distinguished between linear and nonlinear polymers; and integrated natural with synthetic polymers. This lucid presentation permitted those, not previously familiar with polymer science, to take advantage of the organic and physical chemistry already known and to extend this information to the field of polymers and polymerization reactions. It represented a giant step towards encouraging fellow chemists to enter the polymer arena and to contribute to the newly developing science of the macromolecule.

No industrial or academic research group of any size develops on the shoulders of a single individual. Carothers' group included J.A. Arvin, Gerard Berchet, Donald Coffman, Arnold Collins, Martin Cupery, G.L. Dorough, Harry Dykstra, Paul Flory, Ralph Jacobsen, Glen Jones, Willard McEwen, W.R. Peterson, George Rigby,

E.W. Spanagel, Howard Starkweather, Frank van Natta, and Julian Hill. Julian Hill worked extensively with Carothers and the two enjoyed a number of joint ventures, among which was the industrially important process of cold drawing.

The group had constructed a molecular still and chose the polyester derived from propylene glycol and hexadecamethylene dicarboxylic acid for the first experiment. A waxy polyester was prepared and heated in the molecular still for an extended period. The product was now tough with a chain length of about forty repeating units. Filaments of the ester could easily be drawn "taffy-like" from the melted product. When stress was applied to the solidified filaments, the thin opaque cylinder of filament separated into two sections joined by a thinner section of transparent fiber. As pulling continued, the transparent section grew till the opaque sections disappeared.

X-ray diffraction techniques developed by Herman Mark and others showed that this "cold drawing" was actually bringing about a reorientation of the polymer chains along the axis of the fiber. These fibers were much tougher and exhibited tensile strength like that of silk. They were also more pliable and flexible.

This phenomenon of "cold drawing" general occurs with linear and lightly crosslinked synthetic and natural polymers and today is employed in the production of most textiles, rug piles, fishing lines, tire cords....

The first polymers synthesized had melting points below 100°C and were only moderately stable in water, making these polymers unsuitable for textile applications where washing in hot water is a preferred method of cleaning.

On the basis of some preliminary experiments and on the theoretical grounds, the work turned toward the synthesis of polyamides. The initial polyamides synthesized did exhibit high melting points, but were also unstable at high temperatures. Many man hours were spent varying the nature of R and R^1 in an effort to find a product which had both a high melting point and good thermal stability.

The first polyamide that was mechanically spun from a melt was nylon-9, which had a moderately high melting point of 196°C. Considerable work was also done with nylon-5,10. Nylon-5,10 fibers had strengths greater than silk, a melting point of about 200°C and its fibers stood up to moisture and common solvents such as carbon tetrachloride and chloroform, common dry cleaning solvents of that day. Another polyamide, nylon-6,6 was synthesized and though it decomposed just above its relatively high melting point which was 265°C.

After a number of candidates were synthesized and partially characterized, the question was which nylon should DuPont begin its venture with? The key consideration was the availability of the starting materials. Benzene was considered a prime potential feedstock and since it has six carbon, nylon-6,6, with each reactant containing six carbons, was the nylon chosen. This point of ready abundance of inexpensive feedstock has been a critical consideration time and again in industry selecting one compound over another.

As women's hemlines rose in the 1930's, silk stockings were in great demand, but very expensive. Nylon changed this. It could be woven into the sheer hosiery that women desired and was more durable than silk. The first public sale of nylon hose was in Wilmington, DE, on October 24,

1939. They were so popular that they had to be rationed because of the need for nylon for making parachutes.

Nylon, the first fiber produced from a deliberate search (begun in 1927 with the wooing of Wallace Hume Carothers from Harvard by DuPont), was also the first synthetic (man-made) material whose physical properties equalled or exceeded those of the analogous natural (protein fiber) occurring material. Carothers and Hill succeeded in synthesizing and forming nylon fibers some eight years later. In 1937, it appeared on the market as the bristles of toothbrushes and in 1940 as women's stockings. Today it is produced in many forms for many purposes at an annual rate exceeding eight pounds (4 kg) per person in the USA.

Nylon is a strong fiber with a tensile strength of 4,000 to 6,000 kg/cm^2 which is still only about one-fifth of the ultimate strength it would have if the molecules could be perfectly aligned. About half of the nylon fiber goes into tire cord. The remainder is used in ropes, cords, rugs, fish nets and lines, clothes, thread, undergarments, coats and dresses. It is used as an engineering plastic, i.e., as a substitute for metal in bearings, gears, rollers, and as housings and jackets on electrical wires.

Before nylon could be mass produced for public consumption, scientists needed to find large, inexpensive sources of the reactants, i.e., hexamethylenediamine and adipic acid. The DuPont Company scientists devised a scheme for producing these two reactants from coal, air and water. The process was laborious and later scientists developed procedures to make these two reactants from agricultural byproducts such as rice hulls and corn cobs. There are several chemical outlines relating these synthetic procedures which were devised from specific catalysts and emphasize the use of industrial catalysts in the chemical industry.

In the end, spinnable nylon 6,6 proved easier to manufacture than the polyesters since it did not require the use of a molecular still or extended reaction times. Further, the material balance between the diamine and diacid reactants was found to be conveniently achieved through formation of a 1:1 salt of the two reactants which permitted the ready synthesis of long chains of nylon 6,6. Finally, it was found that the initial degradation of nylon 6,6 was oxygen dependent and could be prevented through the use of ultrapure "Seaford" nitrogen developed for blanketing hot nylon 6,6.

Carothers, Hill and co-workers also began developing procedures for the synthesis of cyclic ring compounds. In his paper "Studies on Polymerization and Ring Formation" Carothers noted that bifunctional reactants produced cyclic products when 5- or 6-membered rings were possible. It was known from the work of Ruzicka on natural musk that some large chains were stable and that failure to obtain macromolecular cyclic compounds was due to the lowered possibility of ends of long chains to find one another.

General methods were developed to synthesize a number of cyclic compounds, the largest being a 30 ring cyclic ester derived from decamethylene octadecanedioate. Such studies contributed to the growing chemical knowledge of steric constraints and stable and unstable structural units. As a side note, natural musk and analogous cyclic compounds often exhibited classical perfume odors which followed the researchers home and had to be explained to their wives.

In the late 1920's a group from the Jackson Laboratory of DuPont was working on the polymerization of acetylene to its dimer and trimer, building on the recently reported results of Father Nieuwland. They hoped to find the basis for a synthetic rubber in the chemistry of the dimer and trimer of acetylene.

A study of the fundamental chemistry of acetylene and its dimer and trimer was undertaken by Carothers' group. One of the first steps was to obtain the trimer in pure form. Arnold Collins was conducting the distillation and had turned off the apparatus for the weekend. When he and Hill arrived the following Monday they found the distilled material had solidified. Almost as a reflex response to its feel, Collins threw it against the bench and it bounced in lively fashion. For some time on, the investigation of the white solid took precedence.

Analysis of the white solid showed a high chlorine content, presumably derived from the catalyst, cuprous chloride, used to form the trimer. Preliminary analytical studies indicated the starting reactant was similar in structure to isoprene, the starting material for synthetic rubber. Because of the structural similarity, the starting material, now known to be 2-chloro-1,3-butadiene, was dubbed chloroprene. In fact, it was derived from the dimer and was found to be readily synthesized from the reaction of the dimer and hydrogen chloride.

Polychloroprene resembled vulcanized rubber in its physical properties but was superior in its resistance to ozone, ordinary oxidation and to most organic solvents. Commercial sale of polychloroprene, under the generic name "neoprene" and trademark "Duprene", began in June, 1982, about two years after its discovery.

The chemistry of the dimer and trimer and a number of new intermediates derived from them were further studied.

RECOGNITION AND CLOSURE

Carothers was an extraordinary leader of his coworkers. Although often unresponsive in larger groups, Carothers was a highly effective communicator with his fellow researchers.

Paul Flory writes:
"It was my great good fortune to be assigned to the group headed by Wallace H. Carothers at DuPont in 1934, where I became the only physical chemist among a select contingent of organic chemists. In retrospect, I realize that this placed me in a uniquely favorable position to observe the usual breadth of the interests of Carothers. He was, of course, a brilliant organic chemist. Although he had little training in physical chemistry, his appreciation of that subject, as well as many others, was not limited by the bounds of his formal background. Such limitations as he may have had in mathematics and in physical aspects of chemistry did not deter him from discussing matters falling in these domains, and others as well that were outside his main experience. He was actually among the first to perceive the possibilities of mathematical methods in polymer science, and he lent his support and encouragement to efforts in this direction. His approach to science was motivated by boundless curiosity it was not fettered by superficial boundaries between specialties."

His reputation spread rapidly. In 1929 he was elected Associate Editor of the Journal of the American Chemical Society; in 1930 he became an editor of Organic Synthesis.

He spoke nationally and internationally. In 1936 he became the first industrial organic chemist to be elected to the National Academy of Sciences. (Again it must be noted that fundamental chemistry was almost exclusively done in the academic community and that Du Pont's decision to begin the Experimental Station was as great an innovation as Ford's assembly line.)

He married Helen Everett Sweetman of Wilmington on February 21, 1936. Mrs. Carothers had a bachelor's degree in chemistry from the University of Delaware in 1933 and was employed in the patent division of the chemistry department of Du Pont from 1933 to 1936. A daughter, Jane, was born November 27, 1937.

Early in life he developed a love of reading. He was broadly educated and had a wide fund of information including current problems; critical analyses of politics, labor and business; as well as music, literature and philosophy.

While he was generous, pleasant and modest, he was deeply emotional. Despite his acceptance in the scientific community, Carothers felt he had not accomplished as much as he could and that he might run out of good ideas. Since early manhood he suffered from periods of depressions. These periods grew more pronounced as he grew older despite counsel and the prodding of friends and medical advisors. His sister Isobel's death in January 1936 was a staggering shock to him.

On April 29, 1937, two days after his forty first birthday, he took potassium cyanide in a hotel room in Philadelphia and ended his brilliant career.

Just two years later, in 1939, Du Pont began producing the nylon he had discovered on a commercial scale. This was the first manmade fiber to compare and excel in properties to analogous materials which were produced by nature.

REFERENCES

R. Adams, National Academy of Sciences (U.S.), <u>20</u>, 293 (1939).

R. Adams, "Collected Papers of W.H. Carothers on High Polymeric Substances," (H. Mark and G. Whitby, eds.), Interscience, NY, 1940.

J.W. Hill, "The Robert A. Welch Foundation Conferences on Chemical Research. American Chemistry - Bicentennial," 1976.

CHAPTER 13

HERMAN F. MARK
FATHER OF POLYMER EDUCATION

ABSTRACT

Herman Mark, who celebrated his 93rd birthday in the spring of 1988 was involved in the science of macromolecules as a contemporary of Hermann Staudinger. Professor Mark is a diplomat, author, and lecturer and responsible for the development of polymer science education throughout the entire world.

It has been said that America is a melting pot for immigrants from all other countries of the world. While it is true that millions of foreigners have emigrated to the US, in many instances the process has been one of catalytic fractionation, rather than one of melting. Hitler was the catalyst which caused the Mark family and a Jewish niece to migrate to the US via Canada and as a result of this fractionation process, the new science of polymers developed in the United States under the guidance of a former professor from the University of Vienna.

Since Dr. Mark was not the world's first polymer chemist, one may ask how he can be called the father of polymer science. The title is based on his many contributions to polymer science education and research.

Because of over 5 years of service as a combat soldier in the Imperial Austrian Army, Herman Mark did not receive his Ph.D. degree until the age of 26. However, he was granted the Ph.D. degree, summa cum laude, after two years of study at the University of Vienna in 1921 and then accepted a position at the University of Berlin.

R. B. Seymour (ed.), Pioneers in Polymer Science, 145–151.
© 1989 by Kluwer Academic Publishers.

Among the many chemists at that University were Emil Fischer, Max Bergmann and Herman Leuchs, who were biopolymer science pioneers and Carl Harries, who was a pioneer elastomer scientist.

Dr. Mark returned to Vienna in August 1922, in order to marry Maria (Mimi) Schramek, a catholic, and then accepted a position at the Kaiser Wilhelm Institute in Berlin-Dahlem, where Fritz Haber served as Director and R.D. Herzog as Codirector. Mark's team leader was Michael Polanyi and the principal project was the study of the molecular structure of cellulose, wool, and silk by the use of x-ray diffraction techniques in collaboration with R. Brill and K. Weininberg.

His first of over 600 scientific publications was one in Berichte on the stable free pentaphenylethyl radical in 1922. His first coauthored book with Dr. Epstein on the structure of wool was followed by a second book on x-ray diffraction techniques.

On the recommendation of Dr. Haber, Mark accepted a position in 1927 as assistant director of IG Farbenindustrie in Ludwigshafen am Rhine, where Professor Kurt H. Meyer served as director. In his attempt to elucidate the molecular structure of naturally occurring polymers, Mark was assisted by twenty professional scientists and by such eminent consultants as Drs. Staudinger and Debye.

In addition to authoring or coauthoring some 80 scientific articles while at IG during the period 1927-1932, Mark coauthored a book on the structure of macromolecules with Meyer, wrote a treatise on the physics and chemistry of cellulose, and served as associate professor of physical chemistry at the University of Karlsruhe. His sons, Hans and Peter, were born in Mannheim in 1929 and 1931. Both received Ph.D. degrees in physics. Hans was Under Secretary of the US Air Force and is now chancellor at the University of Texas at Austin. Prior to his death, Peter was a Professor of Physics at Princeton University.

Their father, who was the son of a Jewish physician, Dr. Herman C. Mark, and a Lutheran mother, Lili Mueller, was born in Vienna in 1895. As a leader of a great medical era of the early 20th Century, the senior Dr. Mark, who embraced lutheranism, and his family were associated with many eminent physicians and scientists including Dr. Chaim Weizmann, the founder of the free state of Israel. The Mark family, along with relatives, who were devoted Zionists, had traveled through Israel on several occasions.

Thus, in recognition of the growing Nazi domination, Mark left Germany in 1932 and accepted a position as director of the first Chemical Institute at the University of Vienna, where he continued his research and teaching in the new field of polymer science. Dr. Rudolf Raff, eminent polyethylene scientist, was one of the first doctoral students.

Mark was author or coauthor of another hundred articles while at Vienna during the period 1932-1938. His institute attracted such visiting scientists as Guilio Natta from Italy and others from all over the world.

By moving from Mannheim to Vienna, the Marks had delayed their eventual migration from Europe by 6 years. Hitler invaded Austria in March, 1938, and Professor Mark was dismissed from the University in April of that year.

The Mark family, with only their clothes hung on coat hangers made from platinum wire, crossed the Swiss border disguised as alpine vacationers and eventually arrived in Canada. The platinum proved much more valuable than the 200 billion marks that he accumulated after World War I. He obtained $400 by selling the coat hangers in Switzerland.

He accepted a position as research manager of the Canadian International Paper Company in Hawkesbury, Ontario, Canada, in 1938. While with the International Paper Company (1938-1940), he started a series of books on high polymers for Interscience Publishing Company. The first book of this series was "Collected Papers of W.H. Carothers," which he coauthored with G.S. Whitby. The second book in the series, "Physical Chemistry of Polymers," was authored by Mark.

In 1940, he accepted an adjunct professorship at Polytechnic Institute in Brooklyn, which provided a teacher's visa and a consultanship with the DuPont company. his first course, "General Polymer Chemistry," attracted few graduate students but attracted auditors such as Emil Ott, Milton Harris, H.M. Spurlin, Calvin Schildknecht, I. Allen, H. Bender, W.J. Hamburger, E.O. Kraemer, and J.D. Sally.

Dr. Mark was appointed head of the Shellac Bureau in 1939 and was promoted to a full professor in 1942. During the World War II years, Mark directed many research projects for the United States government. Among the many notable polymer scientists on his research teams were Drs. Turner Alfrey, Paul Doty, Isidor Fankuchen, W. Hohenstein, Arthur Tobolsky and B.H. Zimm.

After the war, Professor Mark organized the Institute of Polymer Research in a vacant razor blade factory on Jay street in downtown Brooklyn. The principal additional expense for the institute was the cost of printing stationery with the now well known 333 Jay Street address.

With the aid of Dr. E.S. Proskauer, he established the Journal of Polymer Science, which he has continued to edit up to the present time. He also continues to add to the volumes in the High Polymer Series, Polymer Reviews and the Encyclopedia of Polymer Science and Engineering.

The Institute of Polymer Science, which was the first American institute to confer the Ph.D. degree in polymer science, became the incubator for many of the nation's polymer scientists. This list includes Drs. C.G. Overberger, Murray Goodman, H. Frisch, H. Gregor, H.P. Frank, F.R. Eirich, E. Immergut, R.B. Mesrobian, G. Oster, J. Schurz, and H. Morawetz. The list of visiting scientists, which is equally impressive, includes Drs. B. Ranby (Sweden), N. Ogata (Japan), G. Burnett (England), J. Aleman (Spain), A. Liquori (Italy), L. Valtassari (Finland), H. Gibello (France), E. Katchalsky (Israel), S. Palit and M.S. Muthana (India), and E. Oleynik and V. Kabanox (USSR).

Professor Mark was appointed Dean of the Faculty in 1961 and was named Dean Emeritus in 1964 when he retired at the age of 69. His successor as Director of the Institute was Professor Charles G. Overberger.

Herman Mark was one of the selected invitees to the Gibson Island Conferences on polymer science. These conferences were started by Dr. Neil Gordon in a dual attempt to keep the Gibson Island (MD) Country Club solvent and to promote the discussion of the advances in science. These sessions, which are now called the Gordon Conferences, have expanded to over a dozen preparatory schools in New England.

Mark was a skier and mountain climber as well as a tennis and soccer player when in school and continued to play tennis for many years. He is also a linguist, speaking German, Italian, French and English fluently. he also was awarded the Leopold Order and was the most decorated company grade officer in the Austrian army in World War I.

Seventeen universities from ten countries have awarded honorary degrees to Dr. Mark. Among these are Lowell University, Polytechnic Institute of Brooklyn and the University of Vienna. he is a member of over 15 state and national academies, including the U.S. National Academy of Science, the American Academy of Arts and Sciences, the Royal Institute of Great Britain, and the Soviet Academy of Sciences.

He has received more than 30 medals and awards including the Honor Scroll of the American Institute of Chemists (1953), the Nichols Medal of the American Chemical Society (1960), the SPE International Award in Plastics Science and Engineering (1962), the ACS award in Polymer Chemistry (1965), the Chemical Pioneer award of the American Institute of Chemists (1972), the ACS award in Organic Coatings and Plastics (1975) and was named a Pioneer in Polymer Science by Polymer News.

As a combat soldier in the Austrian Army in World War I, Dr. Mark survived almost 5 years of combat in the trenches on all fronts. He was arrested and relieved of his passport before escaping from Austria in 1938. Because of a teacher's visa, he was able to migrate to the United States. He started the world renowned Institute of Polymer Science in an abandoned factory building and after teaching and directing research of over 100 polymer scientists, he has continued, at the age of 93, to serve as an active lecturer, writer and editor. So Dr. Mark is not only the father, he is also the grandfather of polymer science in the USA.

REFERENCES

R.B. Seymour, C.H. Fisher, "Profiles of Eminent American Chemists," Litarvan Enterprises, Sydney, Australia, 1988.

L. Pauling, Chapt. 11 in this book.

J.I. Kroschwitz, "Encyclopedia of Polymer Science and Engineering," Vol. 1. John Wiley, New York, NY, 1985 (This tribute includes a list of 595 publications).

CHAPTER 14

PAULING ON MARK

Linus Pauling
440 Page Mill Road
Palo Alto, CA 94306

Herman Mark is famous for his tremendous contributions to the field of polymer science. He is not so well known for his early work in the field of crystal structure. nevertheless, I think that it was his experience in the crystal structure field that gave him the background that permitted him to make his important contributions to the understanding of polymers. He was, in fact, one of the leading investigators in the field of the use of x-ray diffraction for the determination of the structure of crystals in the years 1923 to 1928, and it was through this work that he developed the feeling for atoms and their interaction with one another that permitted him, later on, to make an effective attack on the problem of the structure and properties of macromolecules.

The first volume of the Strukturbericht, written by P.P. Ewald and C. Hermann, covered the period 1913 to 1928. Herman Mark ties for third in the number of page references on crystal structures reviewed in this compendium. First place is held by the mineralogist Viktor M. Goldsmith, with 89 references, many of them, however, to this great multivolume work Geochemische Verteilungsetze der Elemente. Second place, with 53 references, is held by Wheeler P. Davey, who was diligent in applying the powder technique that had been invented by Hull (independently of Debye) to a large number of elements and simple compounds. The tie for third place, 50 references, it between Mark and R.W.G. Wyckoff.

R. B. Seymour (ed.), Pioneers in Polymer Science, 153–163.
© 1989 by Kluwer Academic Publishers.

In his x-ray work, Mark showed a greater range of interests than the other investigators in the field. Mark and his collaborators studied elements, both metallic and nonmetallic, minerals, inorganic compounds, simple organic compounds, condensed gases, and macromolecular substances, and in addition they studied the physics of x-rays and of the diffraction phenomenon.

Most of Mark's work during this period was done at the Kaiser Wilhelm Institute of the Chemistry of Fibers or the Kaiser Wilhelm Institute for Physical Chemistry and Electrochemistry, Berlin-Dahlen. His collaborators during this six-year period included W. Basche, R. Brill, W. Ehrenberg, K.W. Gonell, C. Gottfried, O. Hassel, E.A. Hauser, J. Hengstenberg, H. Hoffman, H. Kallmann, J.R. Katz, H. Mehner, K.H. Meyer, W. Noethling, E. Pohland, M. Polanyi, P. Robaud, J. Steinback, G. von Susich, Leo Szilard, S. Tolksdorf, K. Weissenberg, and E. Wiegner.

When Mark began his x-ray work, the crystal structure had not yet been determined for any organic compound. At that time the x-ray techniques could be applied, with the greatest prospect of success, in determining the complete structure to crystals with high symmetry, especially cubic crystals. It is not surprising that Roscoe G. Dickinson, who had begun crystal structure work in 1917 at the California Institute of Technology and who was hoping to be the first to determine the structure of an organic crystal, should have selected hexamethylenetetramine, $C_6H_{12}N_4$, one of the few organic compounds that forms cubic crystals, for his investigation, and that Mark himself should have selected the same substance. In January 1923, Dickinson and a student of his, Albert Raymond, published their determination of the structure, and eight months later Mark and Gonell published their determination, with essentially the same results. Dickinson soon went into another field of research, but Mark continued for some years to investigate crystals or organic compounds. In most cased he found it impossible to make a complete determination of the positions of the atoms, but he usually succeeded in drawing some conclusions about the structure of the organic molecules from the x-ray results. With Eisenberg in 1923, he published an account of a preliminary study of urea, followed by studies of other organic compounds, usually made in collaboration with von Susich, Mehner, Hassel, Hengstenberg, or Noethling. The crystals investigated include carbon tetraiodide, tetramethylmethane, tetranitromethane, pentaerythritol, metaldehyde, biphenyl, stilbene, triphenycarbonyl, triphenylmethylbromide-fluorene, phenanthrene, D-glucose, D-fructose, and D-cellobiose.

In 1925 he reported on his work on condensed gases, CO_2, B_2H_6, NH_3, and CS_2. carried out together with Polanyi. His structure for carbon-oxygen distance is incorrect. His and Polandyi reported this bond length to be 1.6A rather than 1.16A. The boron-boron bond length found in their study of diborane, 1.8 to 1.9A, is close to the correct value, 1.77A. His first study of an element was carried out in 1923, with Weissenberg, Gonell, and Wiegner. They reinvestigated orthorhombic sulfur, which had been reported by W.H. Bragg to have a unit cell containing sixteen sulfur atoms. Mark and his collaborators found that each of the edges of the unit had to be multiplied by 2, to give a cell containing 128 atoms. They and Hassel in 1924 reported on their reinvestigation of graphite, which they found to have the structure assigned to it by Hull in 1917, rather than a alternative structure suggested by Debye. They evaluated the parameter as 0 + 0.10. J.D. Bernal, in England, was investigating graphite at the same time, and in the same year he reported similar results: the Hull structure, with the parameter equal to 0 + 0.06.

The structure of white tin had been subjected to investigation first by Bijl and Kolkmeijer in 1918, who reported an incorrect structure. in 1923 Mark and Polanyi carried out a second investigation, and found the correct structure for this tetragonal crystal, a structure involving no variable parameters. In 1924, Mark and Hassel reported the results of their reinvestigation of the structure of bismuth, whose structure had been determined in 1921 by R.W. James, in Manchester, who assigned the value 0.232 + 0.004 to the variable parameter. Hassel and Mark verified the James structure, with the parameter equal to 0.236 + 0.003. The presently accepted value for the parameter is 0.2339.

Mark also extended his interests in 1923 to inorganic crystals, beginning with calomel, mercurous chloride. Calomel had constituted a puzzle for inorganic chemists in the 19th century, in that the position of mercury in the periodic table is such as to lead strongly to the conclusion that mercury is bivalent, as in mercuric chloride, $HgCl_2$. Calomel has the composition that would permit the formula $HgCl$ to be written for it, suggesting univalence. Chemists discovered, however, that calomel and similar compounds of mercury (mercurous mercury) ionize in solution to produce the diatomic cation Hg_2^{++}. In 1923, Mark and his collaborator Weissenberg determined the structure of the crystal, and found it to contain linear molecules Cl-Hg Hg Cl, with the distances indicating that the molecule contains a mercury-mercury covalent bond as well as mercury-chlorine covalent bonds. This was an important contribution to structural inorganic chemistry.

Other inorganic crystals studied by Mark and his collaborators, sometimes leading to complete structure determinations, included strontium chloride, zinc hydroxide, tin tetraiodide, potassium chlorate, potassium permanganate, and ammonium ferrocyanide. Minerals investigated by them include $CaSO_4$ (anhydrite), $BaSO_4$ (barite), $PbSO_4$, Fe_2TiO_5 (pseudobrookite), and three forms of Al_2SiO_5 (cyanite, andalusite, and Sillimanite).

Mark's x-ray work on fibrous macromolecular substances began in 1925, with his publication, together with Katz, of a paper on cellulose. He continued the work on cellulose with Meyer and Susich (1929). In 1929 he and Hauser published a report of their studies of rubber. He had developed excellent ideas about the nature of rubber and the explanation of its extensibility and elasticity. I remember that when I visited him in Ludwigshafen in the summer of 1930 both he and I took pleasure in a demonstration that he showed me. He took a large piece of unvulcanized rubber, and stretched it to twice its length. When it was released, it contracted to its original length. He then stretched the rubber and held it under a cold-water faucet, so that it was cooled in the stream of water. On being released, it remained in the stretched form, which had crystallized. He discussed the part played by the increased entropy of the contracted form in the extensibility of rubber, much to my edification.

During this period he had already developed an interest in the fibrous proteins. R.O. Herzog and W. Jancke had made moderately good x-ray fiber diagrams of silk fibroin in 1920, and in 1923 Brill had assigned indices to a about twenty spots in the fiber diagram in terms of an orthagonal unit with identity distances of 7.0A along the fiber axis and 9.3 and 10.4A perpendicular to it. Meyer and Mark in 1932 then discussed the structure more definitely, on the basis of Brill's x-ray data. They suggested, as Brill had earlier, that the polypeptide chains extend along the fiber axis, and contain glycine residues alternating with amino acid residues with larger side chains, principally alanine and serine. Four polypeptide chains, extending along the fiber axis, would pass through the unit cell, with one glycine residue and one alanine residue of each chain in the cell. They also suggested that the peptide chains are strongly attracted to one another by forces between the CO groups and the NH groups of adjacent chains.

None of their conclusions could be drawn from the analysis of the x-ray diagrams alone; instead, they were the result primarily of considerations of the length of chemical bonds, the bond angles, and the nature of intermolecular forces.   Although the nature of the hydrogen bond was pretty well understood by 1932, twelve years after the first important paper on the hydrogen bond, by W.M. Latimer and W.H. Rodebush of the University of California in Berkeley, had been published, Mark, in common with most other European scientists, still had little knowledge about this important structural feature, and accordingly he and Meyer could not make their suggestions about the interaction of the CO and NH groups precise.

In fact, even after the later refinement of experimental techniques, including the preparation of doubly-oriented fibers of silk fibroin by the stretching and rolling of silkworm gut and the discovery of the Petterson diagram and other increasingly powerful methods of interpreting x-ray data, it was still not possible to derive the structure of silk fibroin except through the detailed application of principles of structural chemistry.   By 1948, enough detailed x-ray analyses of the structure of crystals of amino acids, simple peptides, and other simple substances related to polypeptide chains, all carried out by Robert B. Corey and his associates in the California Institute of Technology, had provided precise information from which the interatomic distances and bond angles in polypeptide chains could be predicted with high reliability, about 0.01A in bond lengths and 1° in bond angles.   In particular, these observations verified the predicted planarity of the peptide group and the importance of the formation of maximum possible number of N-H...O=C hydrogen bonds.

The application of these structural principles and the use of accurate values for interatomic distances and bond angles permitted the exact description of several possible configuration of the polypeptide chain, the alpha helix and the two pleated sheets. In particular, it was found that acceptable sheet structures of polypeptide chains could not be formed by fully extended polypeptide chains; instead, the chains need to be contracted somewhat, and stiffened in the direction perpendicular to the fiber axis and the material hydrogen bonds. The predicted length of the two-residue unit of a completely extended polypeptide chain is 7.23A, that for the antiparallel chain pleated sheet is 7.00A, and that for the parallel chain pleated sheet is 6.6A.

This close agreement of the predicted value 7.00A and the experimental value 6.97A for the fiber axis unit length in silk fibroin is strong indication that the structure is based upon antiparallel chain pleated sheets. Moreover, the length of the a axis is calculated to be 9.5A for the antiparallel chain pleated sheet, which is in good agreement with the observed value 9.4A. In 1955, R.E. Marsh, R.B. Corey, and L. Pauling showed that the antiparallel chain pleated sheet structure accounts satisfactorily for the intensities of the x-ray diffraction maxima of silk fibroin. The nature of the structure of silk fibroin suggested by Meyer and Mark in 1932 was thus completely substantiated, and somewhat refined, twenty-three years later.

In the meantime, as is well known, Herman Mark had continued to apply his understanding of the basic structure of fibrous macromolecules in many important ways. Work on the structure of crystals and fibers was not the only way in which Mark made use of x-rays. With several collaborators, he reported the results of a number of significant investigations of the physics of x-rays in 1926 and 1927. With Ehrenberg, he reported studies of the index of refraction of x-rays, and with Leo Szilard, studies verifying the linear polarization of x-rays scattered from electrons at 90°. An investigation of the width of x-ray lines was carried out by Mark and Eherenberg, and Mark and Kallmann reported work ·on the properties of Compton-scattered x-radiation and on the theory of the dispersion and scattering of x-rays.

I cannot conclude this account without mentioning another contribution to science by Herman Mark, not involving x-rays. I believe that this contribution, the discovery of the technique of determining the structure of gas molecules by the diffraction of electrons, constitutes Mark's most important contribution to structural chemistry, one which, moreover, was of great significance in my own development. In 1930, when I visited Herman Mark in Ludwighafen, I learned that he and his young associate R. Wierl, had constructed an apparatus for scattering a beam of electrons from gas molecules and had determined the interatomic distances in carbon tetrachloride and a number of other molecules by analysis of the diffraction pattern (<u>Die Natursissenschagtern</u>, <u>18</u>, 205, 1930, and later papers by Wierl in <u>Phs. Zeit.</u> and <u>Ann. Phys.</u> in 1931 and 1932). The equations describing the diffraction pattern produced by a wave (x-rays or electrons) scattered by a molecule had been derived independently by R. Ehrenfest and Paul Debye in 1915. The electron-diffraction pattern from a molecule, such as carbon tetrachloride, showed a series of concentric rings, with different intensities, and with the radii of the rings, with different intensities depending upon the interatomic distances and the scattering power of the atoms in the molecule. I was overwhelmed by my immediate realization of the significance of this discovery. For several years I, in common with other x-ray crystallographers, had been disappointed by the repeated failures to determine the structures of crystals by application of the known procedures. What was expected to be a simple structure determination often turned out to be impossibly complex.

For example, in 1922 the crystal $K_2Ni_2(SO_4)_3$ which I had made and examined was found to have a structure determined by nineteen parameters, locating the atoms in the cubic cell, and it was not possible, even in 1930, to determine more than perhaps half a dozen parameters from the x-ray intensities. In fact, despite the interest of many x-ray crystallographers, such as J.D. Bernal in London, in amino acids and simple peptides because of their significance for the problem, even as late as 1937, no one had succeeded in locating the atoms in any amino acid crystal or simple peptide.

The simplest amino acid, glycine, would probably require in its structure determination the evaluation of ten parameters, two for each of the five heavy atoms, based on the most optimistic assumption, that the molecule has a plane of symmetry, and the methods had not yet been discovered for solving such a problem. Whereas the investigation of any crystal by x-ray diffraction was a gamble, in that a simple molecule might interact with its neighbors in the crystal in such a way as to make the structure complex, no such complication effect was possible in a gas.

For example, Dickinson in 1923 had found that the unit of structure of tin tetriodide is a cube containing eight molecules, with the atomic positions determined by five parameters, which he succeeded in evaluating. But the Sn_{14} molecule is tetrahedral, with its structure determined by a single parameter, so that one could predict, with confidence, that the investigation of the vapor by the electron diffraction method would surely permit the verification of the tetrahedral structure and the determination of the value of the one parameter, the tin iodine bond length, without trouble.

As the impact of the significance of this discovery burst upon me I could not contain my enthusiasm, which I expressed to Mark - my feeling that it should be possible in a rather short time, perhaps ten years, to obtain a great amount of information about bond lengths and bond angles in many different molecules. I asked Mark if he and Wierl were planning to continue with such a program, and he said that they were not. He added that if I were interested in building an electron diffraction apparatus he would be glad to help, and in fact he gave me the plans of their apparatus.

On my return to Pasadena in September I talked with a new graduate student in California Institute of Technology, Lawrence Brockway, about this project, and he agreed to undertake the construction of the apparatus (with the help and advice of my colleague Professor Richard M. Badger).

During the following twenty-five years, the structures of molecules of 225 different substances were determined by the electron diffraction method in the California Institute of Technology, through the efforts of 56 graduate students and postdoctoral fellows. These studies led to the discovery of several valuable principles of structural chemistry. I continue to have a feeling of gratitude to Herman Mark for his discovery of this important technique and for his generosity to me in connection with it.

Herman Mark is thought of by most chemists as a pioneer in polymer science. I think of him, with affection and admiration, as a pioneer in modern structural chemistry and an important early contributor to its development. I am pleased to report that the organizers of this important symposium on Pioneers in Polymer Science agree whole heartedly with me on this statement.

CHAPTER 15

PAUL J. FLORY
NOBEL LAUREATE AND POLYMER SCIENTIST

ABSTRACT

Paul Flory spent about equal time as a research scientist in industry and academia but his preference was for the latter. He was born in Sterling, IL in 1910 and after making important contributions to the science of macromolecules at DuPont, Standard Oil, Goodyear, Mellon Institute and the Universities of Cincinnati and Cornell, migrated to Stanford where he was honored by the receipt of numerous awards including the Nobel Prize in Chemistry. He died at Big Sur, CA in 1985.

INTRODUCTION

In spite of his many contributions to polymer science, only a few biographies have been written on Nobel Laureate Paul John Flory (1)(2). His contributions are cited in many treatises (3)(4) but like Alfrey, Huggins, Mark, Marvel, Merrifield, and Pauling, he is not mentioned in the leading biographical reference on American chemists and chemical engineers (5). Yet, few scientists have contributed more to the macromolecular science in industry and academia than Professor Flory.

Paul was born in Sterling, IL on June 19, 1910. He was the son of Ezra and Martha Brumbaugh Flory. After graduating from Sterling High School in 1927, he enrolled in Manchester College which was supported by the Church of the Brethern. After receiving the B.S. degree in chemistry in 1931, from this school which was 200 miles east of Sterling, he enrolled in graduate school at Ohio State University where he was awarded the Ph.D. degree

165

R. B. Seymour (ed.), Pioneers in Polymer Science, 165–172.
© *1989 by Kluwer Academic Publishers.*

in physical chemistry in 1934. Few chemistry professors had accepted the new science of macromolecules at that time and hence Paul was not well versed in this branch of chemistry when he accepted a position at the DuPont Experimental Station in 1934.

While at DuPont (1934-1938), he was privileged to be associated with Dr. Wallace H. Carothers who was the leading polymer scientist in the U.S. at that time. He also married Emely Catherine Tabor during his tenure in Wilmington. The Florys are parents of Susan, Melinda, and Paul J. Flory.

While at DuPont, Dr. Flory, like his mentor, Dr. Carothers, was involved in the controversies that existed among chemists, such as K.H. Meyer, and H. Staudinger. Carothers agreed with Staudinger that derivatives of cellulose did not form aggregates in dilute solutions. Meyer insisted that micelles of polymers existed in solutions but through an analysis of the thermodynamics of polymeric solutions, Flory proved that Meyer's views were incorrect.

Contrary to the accepted view that the rate of reaction of a polymer condensation would decrease as the reaction progressed and as the viscosity increased, Flory showed that this rate was independent of viscosity (6). He also showed that crosslinking would occur when reactants with more than two function groups were condensed. While at DuPont, Flory demonstrated his versatility by explaining chain propagation of free radical-initiated vinyl polymers (7).

In spite of the university-like atmosphere at the Experimental station and probably because of the untimely death of Dr. Carothers, Paul left DuPont in 1938 and spent the next two years at the University of Cincinnati.

While at Cincinnati, he showed that a poisson distribution of polymer chain lengths was produced when a fixed number of reactants were simultaneously polymerized. Some years later, the term "living polymers" was coined as a label for these macroions.

It was my privilege to know Paul when he was at Cincinnati and I was at Akron. I envied his university association and was surprised when he left Cincinnati to join Standard Oil at Linden, NJ in 1940. Our paths crossed again when I served as a consultant for Rubber Reserve Corp. on the butyl rubber program. However, to my knowledge, Paul did not attempt to correct the difficulties experienced by Standard Oil when its attempts to produce this elastomer at Baton Rouge, LA were unsuccessful. Of course, it is a pleasure for me to report that Dr. John Durland and I were able to solve the production problem and make possible the commercial products of this important elastomer.

While at Standard Oil, Dr. Flory developed a quantitative statistical mechanical theory of macromolecular solutions (8). It was my pleasure to attend the Gibson Island (MD) conferences in the early 1940's and to hear Paul discuss his theory with Maurice Huggins who had arrived at comparable theory, independently. Some may be amazed to learn that polymer chemists, such as Flory and Huggins, would discuss these fundamental breakthroughs at these conferences. However, the other attendees were not permitted to publicize such discussions. Hence, there were many other meritorious discussions at these conferences by polymer chemists of comparable statue.

I surmised that Paul was not particularly happy with his industrial research at Linden, NJ, and was not surprised when he left to accept a position at Goodyear

Tire and Rubber Co. in Akron in 1943. There was little fundamental research at Goodyear during my tenure in the 1930's but Paul did catalyze some basic research there. He showed that the ratio of the rate constants of propagation (k_p) to the square root of the rate constant for chain termination $(k_p 1/2)$ was related to the rate of polymerization.

While at Goodyear, Flory investigated chain transfer which was essential for controlling the chain lengths of GRS rubber (SBR). He also expanded the Flory-Huggins theory to evaluate the free energy requirements for the overlap of polymer coils (9). Flory, like other experts in the field of polymer science, recognized that fundamental principles of macromolecules were not limited to rubber, fiber or plastics but included biopolymers as well.

According to Flory, a globular protein is a three dimensional structure and the transition temperature range which can be monitored by the extinction coefficient at 2950A° represents a change from an essentially one dimensional helix to a random coil (10). Flory presented a statistical mechanical theory of the fusion of polymers which lead to an understanding of the properties and behavior of crystalline polymers (11).

Paul remained at Goodyear until 1948 when he accepted a professorship at Cornell. His tenure at Akron could not be described as a happy one. The Goodyear researchers were topflight but unlike Carothers and Flory, they were industrial scientists whose advancement was dependent on commercially exploitable discoveries and not on advances in fundamental knowledge of polymer science.

He recognized the need for educating his colleagues but did not abandon his interest in rubber elasticity. His Baker Lectures at Cornell were collected in a book,

"Principles of Polymer Chemistry," which was published by the Cornell University Press in 1953. This high level text continues to be the bible of polymer scientists.

In spite of his interest in university research and his less than satisfactory experiences in industry, Professor Flory left Cornell in 1956 to accept a position as Executive Director of Research at Mellon Institute. While at Mellon, Flory formulated a theory for solutions of rod-shaped macromolecules based on a lattice model (12). He left Pittsburgh in 1961 to accept the Jackson-Wood professorship at Stanford where he remained until his death in September 1985.

Paul never taught a course in polymer science at Stanford but he brought additional prestige to this already prestigious institution. His alma mater (Manchester) had already granted him an honorary doctoral degree in 1950 and he was awarded additional honorary degrees by the University of Manchester in England (1960), Ohio State (1970), Weitzman Institute (1976), Indiana University (1977), and Clarkson (1978).

Paul is one of the few American polymer chemist to receive the Nobel Prize (1974). He was also the recipient of the following awards: Baekeland Award (1947), Colwyn Medal (1954), High Polymer Physics Award (1960), Nichols Medal (1947), International SPE Award (1967), Goodyear Medal (1968), Peter Debye Award (1968), Chandler Medal (1970), Carborundum Excellence in Chemistry Award (1971), Gibbs Medal (1973), American Institute of Chemist Pioneer Award (1973), Cresson Award (1973), Priestley Medal (1974), National Medal of Science (1971), ACS Polymer Division Award (1976), Perkin Award (1977) and the Carl-Dietrich-Harries Medal (1978). He was named a pioneer in polymer science by Polymer News (1979).

He was awarded 20 patents by the U.S. Patent Office
and was the author or coauthor of over 300 reports in
scientific journals.   He was a member of the National
Academy of Sciences, the American Academy of Arts and
Sciences and the American Philosophical Society, a fellow
of the American Institute of Chemists, the American
Association for the Advancement of Science and the
American Physical Society. He served as a member of the
board of directors of the American Chemical Society (1959-
1963).

When Paul shook my hand at the First North
American Chemical Congress at Mexico City in 1976, I
was recovering from eye surgery and couldn't see well in
the dimly lit ballroom but Paul graciously informed me of
his identity and discussed our forty years of friendship.
When I received the Honor Scroll from the Louisiana
section of the American Institute of Chemists in 1982,
Paul made the trip to New Orleans to introduce me.   In
this instance, the introducer was equal in importance to
this prestigious award.

He had agreed to participate in the Macromolecular
Secretariat on Advances in Polyolefins at the Chicago ACS
meeting in September 1985.   My cochair, Dr. Tai Cheng,
and I were shocked to learn that he had suffered a heart
attack on the weekend before the meeting, and, of course,
could not be present.   Tai and I dedicated the week-long
symposium and the book of the proceedings (13) to
Professor Flory who will always be remembered by me as
one of the outstanding pioneers of modern polymer science
(14).

REFERENCES

1. G. Gee, Chapter 7 in "Pioneer of Polymer," The Plastics and Rubber Institute, London, England, 1981.

2. R.B. Seymour, Polymer News, 5, 245, 1979.

3. R.B. Seymour, "History of Polymer Science and Technology," Marcel Dekker, New York, NY, 1982.

4. H. Morawetz, "Polymers: The Origins and Growth of a Science," Wiley-Interscience, New York, NY, 1985.

5. W.D. Miles, "American Chemists and Chemical Engineers," American Chemical Society, Washington, D.C., 1976.

6. P.J. Flory, J. Am. Chem. Soc., 58, 1877, 1936.

7. P.J. Flory, J. Am. Chem. Soc., 59, 241, 1937.

8. P.J. Flory, J. Chem. Phys., 10, 51, 1942.

9. P.J. Flory, W.R. Krigbaum, J. Chem. Phys., 18, 1086, 1950.

10. P.J. Flory, "Statistical Mechanics of Chain Molecules," Wiley-Interscience, New York, NY, 1969.

11. P.J. Flory, J. Chem. Phys., 17, 223, 1949.

12. P.J. Flory, Proc. R. Soc. London A., 234, 73, 1956.

13. R.B. Seymour, T. Cheng, "Advances in Polyolefins," Plenum Press, New York, NY, 1988

14. R.B. Seymour, C.H. Fischer, "Profiles of Eminent American Scientists," Litarvan Enterprises Ltd., Sydney, Australia, 1988.

CHAPTER 16

CARL S. MARVEL
THE GRAND OLD GENTLEMAN
OF POLYMER SCIENCE

ABSTRACT

Like Herman Mark and Paul Flory, Carl "Speed" Marvel was involved in polymer science research and education for about a half of a century. One half of his academic life was spent at the University of Illinois and the other half at the University of Arizona where he continued his research activities until his death in 1987. The University of Arizona named its chemistry building after him on the occasion of "Speed" Marvel's 90th birthday.

In the 1940's, the National Association of Manufacturers established a Modern Pioneers Award to recognize Americans who have been awarded over 40 U.S. patents. While he did not receive the NAM Modern Pioneers Award he has received over 20 awards for his outstanding achievements in polymer science.

In addition to receiving the prestigious award of the American Chemical Society at the 9th Biennial Polymer Symposium in 1978, Professor Marvel has received more than 20 awards including honorary doctoral degrees from Illinois Wesleyan University (1946), University of Illinois (1963), and the University of Louvain, Belgium. He has received the ACS awards in Polymer Chemistry (1964) and in Organic Coatings and Plastics (1973).

Other ACS awards include the Nichols medal (1944) and the Willard Gibbs medal (1956). He received the International Award of the Society of Plastics Engineers

R. B. Seymour (ed.), Pioneers in Polymer Science, 173–176.
© *1989 by Kluwer Academic Publishers.*

(1964), the Perkin award of the Society of Chemical Industry, the Pioneer award (1967) and the Gold Medal award (1955) of the American Institute of Chemists. He was named a pioneer in polymer science by Polymer News in 1978.

He has been a member of the U.S. National Academy of Science since 1938, the American Academy of Arts and Science since 1960, the Philosophical Society, also since 1960, and served as president of the American Chemical Society in 1945. One of the large conference rooms in the American Chemical Society Building in Washington, D.C., is named Marvel Hall.

In addition to publishing "Introduction to the Organic Chemistry of High Polymers" and three other books, Dr. Marvel published almost 500 articles in scientific journals throughout the world. He also served on the editorial board of Macromolecules, Journal of Organic Chemistry, Journal of Organic Chemistry, Journal of the American Chemical Society, and the Journal of Polymer Science.

One may ask, "What was this man's secret for such outstanding productivity?" Part of the answer is that Carl Marvel was born in 1894 and continued to be active during all of the 93 years of his life. Like many outstanding scientists, he was born on a farm in the Midwest (Waynesville, IL) and went to school not far from his birthplace. He received the AB and MS degrees from Illinois Wesleyan University (1915) and AM and Ph.D. degrees from the University of Illinois (1916,1920).

His professional career was also launched close to home at his alma mater, where he was associated with Roger Adams, Oliver Kamm, Wallace Carothers, Butch Hanford and many other eminent chemists at the

University of Illinois where he retired as research professor at the age of 67 in 1961.

He was a professor of chemistry at the University of Arizona and professor emeritus at the University of Illinois since that time. Carl Marvel married Alberta Hughes in 1933. The senior Marvels have two children Mary Catharine and John Thomas Marvel. Dr. John Marvel, who received his Ph.D. degree from MIT, is now a research administrator at Monsanto.

In addition to finding time to raise a family while setting the pace in polymer science, Professor Marvel spent a share of his time outdoors. Included in these general articles are: "The Unusual Feeding Habits of the Cape May Warbler" (1948) and "The Blue Grosbeck in Western Ontario" (1950).

The coauthors of his scientific articles supply an insight into the professional relationships of an Illinois chemist who worked for over 40 years in the only job offered to him and spent over 26 years after retirement in his second job. Some of his coauthors were W.A. Noyes (1917), Oliver Kamm (1924), Roger Adams (1920), V. du Vigneaud (1924), W.A. Lazier (1924), W.C. Rose (1929), R.L. Shriner (1935), William J. Sparks (1936), John Cowan (1936), C.G. Overberger (1944), Rudolph Deanin (1947), G.E. Inskeep (1948), John Stille (1956), James Economy (1956), William DePierri (1958), P.V. Bonsignore (1959), J.E. Mulvaney (1960), and Pat Cassidy (1965).

Since Professor Marvel was one of the world's outstanding organic chemists who accidently specialized in polymer chemistry in the late 1920's, his publications covered a wide range of subjects. They ranged from the "Responsibility of American Chemists" to "Thermally Stable Polymers with Aromatic Recurring Units."

Few organic chemists have shown an interest in the chemistry of polymers or what they may call the "sloppy stuff" left over from many reactions. Hence, one may ask why Dr. Marvel has demonstrated a different viewpoint.

According to Marvel, "There are still many organic chemists who have no idea of what a polymer is, why they form, nor do they seem to care."

Fortunately for polymer science and the polymer industry, which comprises a major fraction of the chemical industry, Professor Marvel did care about polymers and knew why they form. Hopefully, other competent organic chemists will continue to follow his example.

REFERENCES

R.B. Seymour, C.H. Fischer, "Profiles of Eminent American Chemists," Litarvan Enterprises, Sydney, Australia, 1988.

P.J.T. Morris, "Polymer Pioneers" Publication No. 5, Center for History of Chemistry, Philadelphia, PA, 1986.

CHAPTER 17

WILLIAM JOSEPH SPARKS
CO-INVENTOR OF BUTYL RUBBER

ABSTRACT

William J. ("Bill") Sparks, disdainful of the tradition synthetic rubbers made from C_4 to C_6 dienes, selected other raw materials for making rubberlike polymers. In 1937 Sparks and colleague R.M. Thomas, chemists with Standard Oil of New Jersey, invented the now-famous butyl rubber, made by copolymerizing isobutylene with small proportions of butadiene or isoprene. While a research supervisor (1939-1940) at USDA's Northern Regional Research Center (Peoria, IL), Spark's rubber expertise helped initiate research that eventually transformed vegetable oils into the elastomer called Norepol. Yielding to attractive offers, Sparks returned in 1940 to the Esso Research & Engineering Co., where he remained (Director of Chemical Research, until his retirement in 1967. Spark's creative genius, which was not limited to polymers, led to numerous patents on various materials, e.g., new fuels, gasoline additives, propellents, encapsulated oxidants, asphalt additives, and food-wrapping films. Born in 1904, and a graduate of Indiana University and the University of Illinois (Ph.D., 1936), Sparks received many honors. He advocated creativity and enhanced status for inventors. Active in professional societies (President, American Chemical Society, 1966), Sparks served both chemistry and chemists with distinction. He and Mrs. Meredith Pleasant Sparks (Ph.D., chemistry, and law degree) parented four children. William J. Sparks died October 23, 1976, in his home in Coral Gables, Florida.

R. B. Seymour (ed.), Pioneers in Polymer Science, 177–192.

INTRODUCTION

One of the miracles of the 20th century was the development - in less than four decades - of a vast and versatile synthetic rubber industry. Life as we know it today would be impossible without the several types of synthetic rubbers (1-28) stemming from this miracle and manufactured annually in large quantities by this new industry. The production of synthetic rubbers in the United States increased from negligible quantities in 1940 to about 3 million tons in 1987, when the domestic consumption of natural rubber was much less (1 million tons) (17). In 1983 world rubber production was 12 million tons, of which 70% was synthetic; United States 1983 consumption of rubber was approximately 2.5 million tons, of which 75% was synthetic (22).

Most of the research leading to today's large-scale manufacture of synthetic rubbers was done in the latter part of the 19th century and the first half of the 20th century. By far, most of the synthetic rubber research was strongly influenced by the earlier research on natural rubber, which had shown that a diene (isoprene obtained from natural rubber by pyrolysis or from turpentine) can be polymerized to produce rubbery materials. It is only natural that the most investigations of synthetic rubber followed the natural rubber experience by emphasizing the polymerization of isoprene, butadiene, and similar dienes.

Robert McKee Thomas and William Joseph Sparks were among the relatively few investigators who defied convention by basing their research on starting materials other than dienes (20,24). In 1937 they copolymerized a monoene (isobutylene) to obtain a novel and highly useful product called butyl rubber. The technical nature and properties of their new rubber have been described in several publication (20, 23, 26). Thomas described the personal aspects of their research when he received the Charles Goodyear Medal in 1969 (23).

Butyl rubber was unique at the time of its invention not only because it is based on a monoene but also because of the severe and unusual conditions of its manufacture (monomers copolymerized in methyl chloride at -100 to -90°C using aluminum chloride catalyst) (13). Because of the early experimental difficulties, the new rubber was sometimes called "futile butyl". These difficulties were the result of extremely high temperatures at the reaction site and were readily solved by better dispersion of initiator and reducing the rate of its addition. When this problem was solved by Drs. Durland and Seymour, they asked Bill if he knew the way to solve the dilemma. He replied, "Yes, but nobody asked for my opinion."

Butyl rubber, unlike the diene rubbers, is substantially free of olefinic linkages after vulcanization and hence resistant to oxidation and weathering. Butyl, one of the major synthetic rubbers, was manufactured in recent years in at least six countries for the total world production (exclusive of USSR manufacture) of approximately 400,000 metric tons annually. In 1979, butyl was the third most important synthetic rubber, volume-wise, in the United States (Table I).

Table I. U.S. Synthetic Rubber Production Thousands of Metric Tons (a)

	1977	1978	1979
Butadiene-styrene	1395	1395	1378
Polybutadiene	361	378	397
Butyl	149	154	194
Polychloroprene	165	161	183
Ethylene-propylene	158	174	175
Nitrile	70	73	75
Polyisoprene	62	---	---
Other (b)	56	138	132
TOTAL	**2418**	**2475**	**2534**

(a) Chemical and Engineering News, June 14, 1982, p. 37.

(b) Excludes polyurethane rubber.

The principal physical properties responsible for butyl's commercial success are low glass transition temperature (about 70°C), high impermeability to common gases including water vapor, and high hysteresis over a useful temperature range (13).

Butyl rubber is more than another important invention. Its coinventors, Thomas and Sparks, introduced not only a new and important material, but also new concepts of marginal and controlled functionality in elastomers. These new concepts were helpful in subsequent elastomer work in many different laboratories. The timing of the butyl rubber development was important, too. Invented in 1937, butyl rubber provided valuable assistance to the United States in World War II.

In addition to being a creative researcher, Sparks (Ph.D. degree, Illinois, 1936) was also a successful administrator, a statesman for science, a distinguished member and officer of professional societies, and a valued consultant to various government and educational institutions (29, 33, 34, 36).

WILLIAM JOSEPH SPARKS: CREATIVE INVENTOR AND STATESMAN FOR SCIENCE

Dr. William J. Sparks, born in Wilkinson, Indiana, on February 26, 1904, lived on the family farm until he started his college education at Indiana University. He worked as an industrial chemist before entering graduate school at the university of Illinois in 1934. in 1935, Dr. Sparks received the Ph.D. degree under the direction of Dr. Carl S. ("Speed") Marvel at the University of Illinois. This may be considered an early indication of his interest in the economic importance of research.

Dr. Sparks, known as "Bill" to his friends, made many wise decisions.   At the age of 18, he made the wise decision to select a college education at Indiana University over the gift of a new Model T Ford.

While a sophomore at Indiana University, he wisely accepted the advice of Professor Frank C. Mathers to change from history to a chemistry major.

A third wise decision was made in 1930, when he persuaded classmate and chemistry major, Meredith Pleasant, to become his wife.

In 1934, he (with Mrs. Sparks) entered the Graduate School at the University of Illinois, doing their thesis work under the direction of the late Dr. Roger Adams.   She received a law degree in 1958, and is conducting a successful business as an attorney specializing in patents and technical matters. She has received many honors and served as President of the National Association of Women Lawyers in 1981-82 (35).

In 1937, only one year after receiving the Ph.D. degree, Dr. Sparks decided to start inventing early instead of waiting until late in his career. The first big invention, really a coinvention, butyl rubber, was with colleague, Robert M. Thomas.

The 1939 decision to leave Esso, now EXXON, for employment with the USDA's Northern Regional Research Center, Peoria, Illinois, was fortunate because he was soon invited by EXXON to return under more attractive conditions.

Dr. Sparks spent most of his career with Esso. He was Director of the Chemical Division from 1946 to 1958. He held the prestigious position of Scientific Advisor from

1958 to 1970. For brief periods he did research also at Sherwin-Williams (1926-29), DuPont (1929-34), and USDA's Northern Regional Research Center (1939-40). He was self-employed in Miami, FL, after his retirement in 1970 from EXXON.

Although Dr. Sparks devoted much time to managing research and serving many professional organizations, various Federal agencies, and several educational institutions, he still found time for research and invention. He authored a considerable number of papers on research and professionalism. His name appeared on some 145 patents as inventor or coinventor. Dr. Sparks' first patent issued in 1934, two years before he received the Ph.D. degree. His last patent issued in 1978, two years after his death. Versatility was displayed by his inventions, which described various processes and many types of products, from rubber to propellents and paving compositions.

His patents (many with coinventors) were concerned with many subjects, including a leavening process, refining hydrocarbons, plastics, manufacture of alkyl nitrates, lubricants, additives, corrosion-proof liners, diesel fuel, alcohol-gasoline compositions. plasticizers, stabilized polymers, paving compositions, seed treatment, rocket propellent, and encapsulated oxidants.

Butyl rubber was and is enormously successful both technologically and in the business sense. But other Sparks inventions and coinventions also were commercially successful. Examples are styrene-isobutylene copolymers as coating for paper and paper milk bottles, styrene-isoprene copolymers as artificial leather, colored asphalt paving materials, oxo alcohols, and oxo ester plasticizers. During his brief period of employment as a research supervisor at USDA's Northern Regional Research Center, Peoria, Illinois, Dr. Sparks initiated work on dimer acids and polymers made from vegetable oil. Continuation of this work by NRRC scientists led to the development of the elastomer called Norepol (30).

It is well known, of course, that Bill Sparks was a great scientist and inventor. But probably few are aware of the extent and importance of his dedicated services to chemists, the chemical profession, and professional and other organizations.

Another great chemist, Dmitri Mendeleev, once wrote, "The greater man's natural gifts, the greater his responsibility to society." This general philosophy characterized the career of Bill Sparks.

Unlike some scientists, Dr. Sparks did not think society should support him and his personal preferences in research. Instead, he believed research should benefit society and pay for itself through such benefits.

His spirit of service is reflected in the titles of some of his talks and articles:

"Good Chemists Never Quit," The Chemist, 38, 357-365 (1961).

"Inventions Vital Third Dimension of Science," The Chemist, 42, 107-108 (1964).

"Creativity, Competition, and Cooperation: The Combination to Prosperity," The Chemist, 44, 65 (1967).

"Scientists and the Development of a Social Conscience," The Chemist, 47, 182-187 (1970).

Dr. Sparks' conviction that science should serve society is indicated also by some of his published comments:

"Science without a purpose is an art without responsibility."

"The scientific profession has become much larger than medicine, law, or the clergy.   Yet many young scientists are not taught by their professor to feel an obligation to society in their work."

"If he (the scientist) knows the world he lives in, he will know how to serve it."

"The image of science is going to suffer if people don't get any benefit from the money that is being spent on it."

Dr. Spark's services to science and technology included active participation in many professional organizations. These included the American Chemical Society, American Institute of Chemists, American Institute of Chemical Engineers, Society of the Chemical Industry, American Association for the Advancement of Science, Association of Research Directors, National Academy of Engineering, Armed Forces Chemical Association, American Academy of Achievement, Cosmos Club (Washington), Chemists Club (New York), Sigma Xi, Alpha Chi Sigma, and Phi Lambda Upsilon.

Dr. Sparks gave generously of his time to the American Chemical Society and the American Institute of Chemists.   He held many positions in these organizations, and received their highest honors (ACS Priestley Medal, 1965; Charles Goodyear Medal, 1963; AIC Gold Medal Award, 1954; Chemical Pioneer Award, 1970; and Honorary Membership, 1954).   He was President of the

American Chemical Society in 1966 and a member of the
AIC Board of Directors, 1960-63.

Dr. Sparks displayed not only efficiency but also
versatility and dedication in serving various organizations.
As a diplomat he headed the US Delegation to the
Stockholm meeting of the International Union off Pure and
Applied Chemistry in 1953. He was an advisor to the US
State Department, the US Department of Agriculture, the
US Army, Rutgers University, and Rensselaer Polytechnic
Institute. He was national chairman of the Division of
Chemistry and Chemical Technology of the National
Research Council, the Scientific Research Society of
America, and the Society of Sigma Xi.

Similar instances of service on the part of Dr. Sparks
include: US State Department Delegate to IUPAC in 1955
and 1957; Member, Governing Board, National Academy of
Science, 1953-55; Policy Committee Member, National
Academy of Engineering; Advisory Committee Member,
International Technological Assistance; Committee
Chairman, Chemical and Biological Warfare, Armed Forces
Chemical Association; and Vice President, Chemists Club
(New York).

Dr. Sparks' achievements as a scientist and inventor
and his dedicated services to hiss profession were rewarded
by many important honors. His first major honor was
received in 1926 from Indiana University when several
organizations conferred their highest honors upon him. He
received honorary degrees and was elected to the National
Academy of Engineering. The Library of Congress
requested an autographed photograph and list of
publications and placed these in its Collection of
Photographs of Famous Scientists (33).

WILLIAM J. SPARKS - HONORS

1926 A.B. degree, with distinction, Indiana University

1954 Gold Medal, American Institute of Chemists

1954 Honorary Member, American Institute of Chemists

1956 Distinguished Alumni Award, Indiana University

1963 Charles Goodyear Medal, American Chemical Society

1964 Perkin Medal, American Section, Society of the Chemical Industry

1965 Priestley Medal, American Chemical Society

1966 President, American Chemical SOciety

1966 Honorary D. Sc. degree, Indiana University

1966 Honorary D. Sc. degree, Michigan Technological Institute

1967 Elected to National Academy of Engineering

1970 Chemical Pioneer Award, American Institute of Chemists

1974 Autographed photograph and list of publications and patents requested by Library of Congress, placed in its Collection of Photographs of Famous Scientists

Dr. Sparks received honors also from his principal employer, EXXON, with the insistence and an attractive offer in 1940 that he return to EXXON employment; appointment to Director of the Chemical Division; gift of a gold watch when he became the inventor or coinventor of 100 patents; and appointment in 1968 to the prestigious position of Scientific Advisor.

Dr. Sparks continued to invent and serve the chemical profession following his retirement from EXXON in 1970. Twelve patents were issued during his post-retirement when he served as National Chairman of the Scientific Research Society of America. Only serious illness and death could stifle his zeal for invention and his dedication to the chemical profession and society.

Dr. Spark's heavy involvement in research and other professional activities left little time for hobbies. Nevertheless, he occasionally enjoyed poker and golf. He was a member of the Chemistry Club in New York, the Cosmos Club in Washington, D.C.; the Echo Lake Country Club in Westfield, NJ; and the Riviera Country Club in Coral Gables, FL. In terms of years, Dr. Sparks lived approximately the usual life span. In terms of activities, accomplishments, and impact upon his profession and society, however, it can be said he lived much longer.

REFERENCES

1.   Billmeyer, F.W., Jr., Textbook of Polymer Chemistry, Interscience Publishers, New York, 1957.

2.   Bolker, H.I., Natural and Synthetic Polymers, Marcel Dekker, In., New York, 1974.

3.   Craver, J.K., and Tess, R.W., Applied Polymer Science, American Chemical Society, Washington, D.C., 1975.

4.   D'Ianni, J.D., Kirk-Othmer Encyclopedia of Chemical Technology, Vol. 21, Interscience Publishers, New York, 1953.

5.   Fisher, C.H., The Chemist, June 1966, p. 193-199.

6.   Fisher, H.L., Chem. and Eng. News, 21, (10), 741-750 (1943).

7.   Flory, P.J., Principles of Polymer Chemistry, Cornell University Press, Ithaca, New York, 1953.

8.   Heilbron, I., ed., Thorpes' Dictionary of Applied Chemistry, 4th ed., Longman, Green & Co., New York, 1950.

9.   Horne, S.E., Gibbs, C.F., and Carlson, E.J., British Patent 827,365 (1954).

10.   Houwink, R., Elastomers and Plastomers, Elsevier Publishing Co., Inc., 1949.

11.   Ihde,A.J., The Development of Modern Chemistry, Harper and Row, New York, 1964.

12.   Lenz, R.W., Organic Chemistry of Macromolecules, Wiley-Interscience, New York, 1967.

13.   McGrath, J.E., Baldwin, F.P., and Schatz, R.H. Kirk-Othmer Encyclopedia of Chemical Technology, 3rd ed., Vol. 8, John Wiley & Sons, Inc., 1979.

14.   Meyer, K.H., Natural and Synthetic High Polymers, 2nd ed., Interscience Publishers, New York, 1950.

15.   Morton, M., Rubber Technology, 2nd ed., Van Nostrand Reinhold Co., New York, 1973.

16.   Murphy, W.J., Chem. & Eng. News, 21, (11), 864-877 (1943).

17.   Powers, P.O., Synthetic Resins and Rubbers, John Wiley & Sons, Inc., New York, 1943.

18.   Semon, W.L., Chem. & Eng. News, 21, (19), 1613-1619 (1943).

19.   Seymour, R.B., History of Polymer Science and Technology, Marcel Dekker, Inc., New York, 1982.

20.   Seymour, R.B., and Carraher, C.E., Jr., Polymer
        Chemistry.   An Introduction, 2nd ed., Marcel
        Dekker, Inc., New York, 1988.

21.   Skolnik, H., and Reese, K.M., A Century of
        Chemistry, American Chemical Society, Washington,
        D.C., 1976.

22.   Sparks, W.J., Charles Goodyear Medalist Address,
        Rubber World, 1963, No. 1, p. 74.

23.   Stahl, G.A., Polymer Science Overview: A Tribute
        to Herman F. Mark, American Chemical Society,
        Washington, D.C., 1981.

24.   Stintson, S.C., Chem. & Eng. News, April 25, 1983,
        p. 23-40.

25.   Thomas, R.B., Rubber Chem. & Technol., 42, (4)
        G90-G96 (1969).

26.   Thomas, R.M., and Sparks, W.J., Chapter 24,
        Synthetic Rubber, G.S. Whitby, ed., John Wiley &
        Sons, Inc., New York, 1954.

27.   Wakeman, R.L., The Chemistry of Commercial
        Plastics, Reinhold Publishing Corp., New York,
        1947.

28.   Whitby, G.S., Synthetic Rubber, John Wiley & Sons,
        Inc., New York, 1954.

29.   Cattell Press, American Men and Women of Science,
        Chemistry 1977, R.R.Bowker Co., New York, 1977.

30.   Cowan, J.C., Ault, W.C., an Teeter, H.M., Ind. Eng.
        Chem., 38, 1138-44 (1946); and 41, 1647-52 (1949).

31. Peterson, W.H., Rubber Chem. & Technol., 42, (4), G87-G89 (1969).

32. Reese, K.M., Chem. & Eng. News, Nov. 14, 1983, p. 70; Nov. 28, 1983, p. 56; and Dec. 19, 1983, p. 78.

33. Sparks, M.P., Private Communication.

34. Who's Who in America, 38th ed., Marquis Who's Who, Inc., Chicago, 1975.

35. Women Lawyers Journal, 68, (1), (1982).

36. Fisher, C.H., William J. Sparks, Creative Inventor in Chemicals and Polymers, Division of the History of Chemistry, American Chemical Society, Aug. 30, 1983, Washington, D.C.

37. Considine, D.M., ed., Van Nostrand's Scientific Encyclopedia, 5th ed., Van Nostrand Reinhold Co., New York, 1976, p. 871-3.

38. Engineers Joint Council, New York, 1970, Engineers for Distinction.

39. Malcomson, R.W., CHEMTECH, May 1983, p. 286-292.

40. Baum, V., The Weeping Wood, Doubleday, Doran & Co., Garden City, New York, 1943.

41. Who's Who in America, Vol. 35, 1968-69, Marquis Who's Who, Inc., Chicago, IL.

42. T.H. Rogers, Jr., Natural Rubber, Vol. 11, p. 810-826, Kirk-Othmer Encyclopedia of Chemical

Technology, Interscience Publishers, Inc., New York,
1953.

43.   Who Was Who, Marquis Who's Who, Chicago, IL.

CHAPTER 18

ROBERT M. THOMAS
CO-INVENTOR OF BUTYL RUBBER

ABSTRACT

Robert McKee Thomas, the senior patentee of butyl rubber worked for Standard Oil Co. of New Jersey during his entire career. He was awarded over 75 patents by the US Patent Office and directed the research of several other notable polymer scientists.

INTRODUCTION

If one were to list the ten most important developments in modern science, he would include (1) the discovery of the vulcanization process for rubber and (2) the development of polymerization techniques for converting small molecules into macromolecules. Hence, it is important to note that Robert McKee Thomas, in cooperation with Dr. William Sparks utilized both of these important concepts in their discovery of butyl rubber.

One of the disadvantages of vulcanized soft natural rubber is its lack of resistance to deterioration and to reactants which attack the residual double bonds present. Thomas and Sparks tailor-made a diene copolymer (butyl rubber) which contained enough unsaturation for crosslinking but was more resistant to deterioration than hevea rubber and the crosslinked (vulcanized) elastomer contained few double bonds.

While this invention may appear obvious, it should be pointed out that polyisobutylene hade been produced by cationic polymerization by Butlerov and Goryainov in 1873 and nature has been making polyisoprene for thousands of

R. B. Seymour (ed.), Pioneers in Polymer Science, 193–195.
© 1989 by Kluwer Academic Publishers.

years. However, I.G. Farbenindustrie was unsuccessful in its attempt to produce the copolymer, now called butyl rubber.

Thomas reduced the cold flow of polyisobutylene by blending it with about 10% of natural rubber in the early 1930's. However, the copolymer of isobutylene and isoprene was not produced until the late 1930's, after many years of research efforts. This important copolymer was patented by Thomas and Sparks [U.S. Pat. 2,356,128 (1940)]. This new elastomer was also described in articles by R.M. Thomas and co-workers in 1940-41 in several journals including Ind. Eng. Chem., 32, 1283 (1940).

I had become acquainted with butyl rubber through conversations with Drs. Paul Flory and William Sparks at the Gibson Island (MD) AAAS conferences in the late 1930's and early 1940's. Later Dr. John Durland and I were employed as consultants by Rubber Reserve on the first commercial production of this elastomer in 1943 at Baton Rouge, LA.

John and I met Bob Thomas while we were in Baton Rouge. Both of us maintained that we solved the low molecular weight problem, independently, by suggesting atomization of the catalyst and better dispersion. However, we learned later that this innovation had already been patented by Thomas and Lightbrown (U.S. Pat. 2,275,893). Had the teachings of this patent been used, this elastomer would never have been called "Futile butyl".

In addition to taking the butyl rubber project from conception to commercialization on a large scale, Bob Thomas worked with Baldwin in development of Chlorobutyl rubber (U.S. Pat. 2,926,718) and with Minkler on the development of butyl terpolymers.

Robert McKee Thomas was born in El Paso, TX, on June 21, 1908. In 1918, his family moved to Charlottesville, VA, where Bob graduated from high school in 1925. He majored in chemistry at Virginia Polytechnical Institute and State University and received his B.S. degree in 1929.

After graduation, he accepted a position with Standard Oil Company of New Jersey and retired as Senior Research Associate at this company after 36 years of service in 1965. During his career at Standard Oil (now EXXON Co.), he was awarded more than 75 patents by the U.S. Patent Office and published more than 20 articles in scientific journals. Dr. J.P. Kennedy, another pioneer in Polymer Science served as his co-author on a number of these publications.

Bob was recognized as a fifty year member of the American Chemical Society. He received the Charles Goodyear medal in 1969 and his talk on "Early History of Butyl Rubber" was published in Rubber Chemistry and Technology, 42 (4) G90 (1969). He was named a Pioneer in Polymer Science by Polymer News just prior to his death in 1986.

Bob married Margaret McMahon in 1931 and after her death, he married Barbara Lee Jarvis in 1972. He is the father of two sons. Stuart is a captain with Eastern Airlines and Blaine is a Professor of English at Lamar State University.

The R.M. Thomas family moved from Linden, New Jersey to Baton Rouge during WWII and back to New Jersey where he was head of the Rubber Laboratories in the late 1940's. He returned to Baton Rouge in 1965 but in 1975 moved to the vicinity of Kamuela, HI where he died in 1986.

CHAPTER 19

MAURICE LOYAL HUGGINS
A PIONEER IN SOLUTION THEORY

ABSTRACT

While Maurice Huggins spent much of his career in industry, he was more interested in basic research than in practical investigations. his major contributions to polymer science were related to entropy of mixing and the viscosity of dilute solutions. He died in California in December 1982.

INTRODUCTION

The American Chemical Society Division of Polymer Chemistry was preceded by a High Polymer Forum. Dr. Maurice Huggins was an advocate and first chairman of this forum prior to its attainment of division status in the ACS in the early 1950's.

Like several other Pioneers in Polymer Science, Maury Huggins never strayed far from home, prior to receiving the Ph.D. degree, and he continued his scientific endeavors while in his 80's. He published more than 50 scientific papers after his formal retirement from Stanford Research Institute (SRI) at the age of 70. He was awarded four patents by the U.S. Patent office and published almost 200 scientific articles before that time. He is author of a book on "Physical Chemistry of High Polymers."

This Pioneer in Polymer Science was born in Berkeley, California, in 1897. He received the AB, BS, MS and Ph.D. degrees from the University of California at Berkeley in 1919, 1919, 1920, and 1922, respectively. He held fellowships at his alma mater (1922), at Harvard

R. B. Seymour (ed.), Pioneers in Polymer Science, 197–199.
© 1989 by Kluwer Academic Publishers.

(1922-23) and at California Institute of Technology (1923-25). He served as an instructor and assistant professor at Stanford University (1925-1933) and as a Johnston Scholar and associate professor at Johns Hopkins University (1933-36).

He spent 14 years as a research associate at Eastman Kodak Company before returning to Palo Alto as a senior research scientist at SRI where he retired in 1967. While on a 10 month leave of absence from Eastman, he served as a Fullbright lecturer and visiting professor at Osaka and Kyoto Universities. He was associated with the Arcadia Institute for Scientific Research, without salary, for eight years after his retirement from SRI. Considerable information on Dr. Huggins achievements may be found in a preface, in Macromolecules authored by Professor Stockmayer in honor of this pioneer's 80th birthday.

The lattice theory of entropy of mixing, derived independently and contemporaneously by Huggins and Flory and the "Huggins constant 'K'", relating the concentration dependence of the viscosity of dilute polymer solutions are listed among Huggins' major contributions to polymer science. He developed new procedures for more quantitative productions of solubility.

Dr. Huggins served on the editorial boards of Macromolecules, Journal of Catalysis, Journal of Chemical Physics, Journal of Polymer Science and the British Polymer. He also served as chairman of the Rochester Section and councilor of the Santa Clara Valley Section of the ACS, president of the Rochester Section of Sigma XI, chairman of the Rochester Section of the Federation of American Scientists, the American Physical Society, The American Crystallographic Association, the Gibson Island and Gordon Conferences, the IUPAC Committee on

Macromolecules   and   National   Research   Council
Committees.

He received the Frank Forrest Award of the American
Ceramic Society, the A. Cressey Morrison Prize of the New
York Academy of Science, was the recipient of the ACS
Division of Organic Plastics and Coatings (Borden) Award
in 1980 and was named a pioneer in polymer science by
Polymer News.   Previous recipients of the ACS-ORPC
biennial award were Dr. Paul Flory (1976) and Carl S.
Marvel (1978) who were also Pioneers in Polymer Science.
Maurice   was   awarded   honorary   D.Sc.   degrees   by
Technische   Universitat   Clausthal   and   Kent   State
University.   Dr. Huggins received the Dow Chemical
Polymer Award sponsored by the Polymer Division of ACS
at the U.S./Japan Polymer Symposium at Palm Springs,
California, November 1980.

When questioned by a panel consisting of Drs. William
J. Bailey (University of Maryland), F.R. Mayo (SRI
International), N. Ogato (Sophia University of Tokyo), and
R.B. Seymour (University of Southern Mississippi), the
awardee discussed his role in the formation of the Polymer
Division and the IUPAC polymer group. He also discussed
the development of the Huggins-Flory lattice theory of
entropy of mixing, the "Huggins constant 'K'" relating to
the dependence of concentrations on the viscosity of dilute
polymer solutions, the development of the concept of
hydrogen bonding and the three stack structure of
collagen.

Maurice was married in 1928.   The Huggins were
parents of two children. Their son Robert, is a professor
of material science at Stanford University.   Maurice
Huggins died at Palo Alto in 1982.

CHAPTER 20

KARL ZIEGLER
FATHER OF HIGH DENSITY POLYETHYLENE

ABSTRACT

Karl Ziegler who shared the Nobel Prize with Giulio Natta, was a professor of chemistry at the Universities of Heidelberg and Halle before being named director of the Max Planck Institut. His interest in organometallic compounds led to the discovery of the "Ziegler catalyst" which was used to produce high density polyethylene.

On his 70th birthday, Karl Ziegler established the "Ziegler Fund" for the support of research at the Max Planck Institut by writing a check for 40 million marks which was the equivalent of 10 millions dollars in American currency. Such an endowment might have been conventional for a son of wealthy parents or an executive but Karl was the son of a Lutheran minister who earned less than a million marks in his entire lifetime and Karl was a chemistry professor.

Actually, the annual income of the minister's son was not much greater than that of his father until he reached the age of 45. His wife "Frau Professor" raised chickens in order to feed the Ziegler family. The financial turning point in Ziegler's life came when he accepted a position as director of the Max Planck Institut for Coal Research at Mulheim in 1943. He insisted that he would have complete freedom to choose and publish the results of research projects regardless of their relevance to research sponsored by coal companies. There were no restrictions on outside income from royalties from research not related to coal and hence,the income from his polyolefin patents

R. B. Seymour (ed.), Pioneers in Polymer Science, 201–206.

were paid to Dr. Ziegler. These royalties made Karl one of the richest men in Germany.

Karl was born in the parsonage at Helso which is near Kassel, Germany on November 26, 1898. In 1912, his father moved the family to Marburg, near the University where the senior Ziegler had studied theology before being ordained in the Lutheran Church. Karl attended the Real Gymnasium and won the von Behring prize for being the top student during his senior year.

He entered Marburg University with advanced standing so that he started as a second year student. He was drafted in the German army in 1918 but returned to the University after World War I and received the Ph.D. degree in chemistry in 1920. His major professor was Karl von Auwers who insisted that Ziegler conduct research different from that of his major professor after graduation.

Karl married Maria Kurtz in 1921. Their first child was born in 1922 another child was born later. Their daughter became a medical doctor and their son became a physicist who worked with his father at the Institut.

Karl completed the requirements for his habilitation and was made a privatdocent at Marburg University. He joined the faculty at the University of Frankfurt in 1925 and left the following year to accept a position at the University of Heidelberg where he remained until 1936. He accepted a position as head of the Chemical Institut at the University of Halle where he remained until 1943.

The Max Planck Institut which Karl joined reluctantly as director in 1943 was one of several Institutes originally sponsored by the Kaiser under the name of the Kaiser Wilhelm Institut fur Chemie in 1912. The name of the

Institut at Mulheim was changed to the Max Planck Institut fur Kohlenforshung after World War I.

The original director of the Institut was Franz Fischer who was a cousin of Emil Fischer and the inventor of the Fischer-Tropsch process for the production of hydrocarbons, including gasoline, from coal.

The governing body of the Max Planck Institut learned of Karl Ziegler through his publications after the retirement of Franz Fischer. Karl had followed the work of nobel laureate Grignard on metal alkyls, and investigated free radicals, polymerized butadiene, synthesized large ring compounds and had even investigated cantharidin (from "spanish fly").

One of the early awards received by Ziegler was the commemorative medal of the Society of German Chemists which was awarded to him in 1935 as the result of publishing 15 papers on polynomial ring systems.

Leopold Ruzicka, a student of Hermann Staudinger, received the Nobel Prize in 1939 for disproving Baeyer's strain theory by demonstrating that the active components of musk and civet were cyclic ketones with 16 and 17 carbon atoms in the rings. Ziegler employed the Ruggli high dilution principle in the condensation of dicyanoalkanes in 1934 to obtain high yields of these and other alicyclic ketones with 14 to 33 carbon atoms in the rings.

In 1929, Ziegler showed that an organic metallic compound was the active species in the sodium-initiated polymerization of butadiene. The polymerization reaction used to make synthetic rubber during World War I had been discovered by Harries and Matthews and Strange in 1910.

His interest in the Grignard organomagnesium compounds and the adduct formed by the addition of sodium to butadiene led to the synthesis of lithiumalkyls by the reaction of lithium with mercuryalkyls. Since lithium was a scarce and expensive metal, it was fortunate that his institut had been selected as the repository for Germany's entire supply of Lithium during World War II.

Because the reaction of lithium hydride with ethylene was too slow, he substituted the more reactive lithium aluminum hydride. Subsequently, he found that triethylaluminum added to ethylene produced a higher molecular weight trialkylaluminum. This reaction which was investigated by his former student H. Gellert was appropriately called the aufbau or "build up" reaction.

Eventually, the product of the aufbau reaction added more ethylene molecules and a verdangung (displacement) reaction occurred which produced a mixture of linear alkanes. Ziegler applied for patents, published and discussed his results in frequent lectures.

Otto Bayer, research director for the Bayer Company, ridiculed the Ziegler reactions which he labeled "Mulheimer chemie" which would be pronounced in France as Mulleimer meaning garbage can. Bayer missed an opportunity to obtain a license which would have led to the exclusive production of HDPE in Germany.

Subsequently, 1-butene was obtained exclusively in the aufbau reaction. Further investigations by Ziegler and coworkers E. Holzkamp, H. Briel, and H. Martin showed that a trace of nickel from the stainless steel autoclave was responsible for the change in the reaction product. An investigation of salts of other related metals in 1953 showed that elimination of butene did not occur in the presence of titanium salts but that the aufbau reaction was actually accelerated in the presence of these metal salts.

The use of the successful "Ziegler Catalyst" i.e. titanium chloride and triethylaluminum for producing linear polyethylene (HDPE) at ordinary temperature and pressure was licensed by Ziegler to Petrochemicals LTD in England (now owned by Shell), Montecatini (now Montedison) in Italy, Farbwerke Hoechst and Hercules in the U.S. Hercules and Stauffer formed a company, Texas Alkyls for the production of Ziegler catalysts.

In spite of the immense wealth acquired by Ziegler, he missed the opportunity to acquire even greater wealth by not including propylene with ethylene in his patent application for high density polyethylene. He acted as his own patent attorney for many years and his exclusive licenses were broad and written informally. Hercules procrastinated and let its option to license elapse but Ziegler reluctantly signed an exclusive licensing agreement with Hercules later after this firm underwrote a lecture tour to the U.S.

Karl Ziegler who was both fortunate in his research and in his licensing agreements, was a joint recipient of the Nobel Prize with Giulio Natta in 1963. A Russian acquaintance described him as "the last of the alchemists who researched aluminum and turned it into gold."

Karl was an avid traveller and mountain climber who climbed the Matterhorn in 1952. He and Maria, who celebrated their golden wedding anniversary in 1971, were also interested in art and music.

In the company of one of their ten grandchildren, the senior Zieglers took a cruise to observe the total eclipse of the sun at Cape Verde Islands in 1973. Shortly after his return and after attending a Saturday Colloquium at the Institut, Karl suffered a severe heart attack. He died at Mulheim on August 12, 1973. in addition to being a grandfather of 10 grandchildren, he was also the father of all polyolefins and synthetic elastomers produced by the Ziegler catalyst. This and modifications of this catalyst are now used to produce over 20 million tons of polymers annually.

REFERENCES

P.J.T. Morris, "Polymer Pioneers," Center for History of Chemistry, Publication No. 5, Philadelphia, PA, 1986.

CHAPTER 21

GIULIO NATTA
A PIONEER IN POLYPROPYLENE

ABSTRACT

Giulio Natta, who shared the Nobel Prize in chemistry with Karl Ziegler, was an Italian crystallographer who used the "Ziegler catalyst" to produce polypropylene. His wife, Rosita, coined the tacticity terms after Giulio had shown that stereoregular polymers could be produced by the polymerization of alpha substituted ethylenes.

Young polymer chemists who aspire to become millionaires, like Karl Ziegler and Giulio Natta, should be forewarned that neither of these inventors was educated as a polymer scientist. Ziegler was essentially a physical organic chemist who learned some polymer chemistry from Hermann Staudinger and Natta was a crystallographer who read Flory's "Principles of Polymer Chemistry." Both learned enough about macromolecules to conduct their experiments and obtain patents on polyethylene and polypropylene.

Polyethylene had been produced by von Peckmann in the 19th century by the decomposition of diazomethane. Polypropylidene ($-CH-)(C_2H_5$) isomeric with polypropylene had also been obtained by the decomposition of diazopropane. Later, low melting, amorphous polypropylene was produced by the ionic polymerization of propylene but the softening point of this product was too low to consider its use as a plastic.

However, a solid polypropylene which was useful both as a plastic and a fiber was produced by Giulio Natta and several other scientists in the mid 1950's. Based on

R. B. Seymour (ed.), Pioneers in Polymer Science, 207–212.
© 1989 by Kluwer Academic Publishers.

verifiable dates of conception of the invention, the U.S. Patent Office finally agreed, after an extremely long and expensive patent suit, that P. Hogan and R. Banks of Phillips Petroleum Company were the inventors of crystalline polypropylene but chemists at DuPont, Standard Oil of Indiana, Standard Oil of New Jersey, Shell, Hercules and the Max Planck Institut were all contemporary inventors.

Karl Ziegler of the Max Planck Institut patented the polymerization of ethylene at ordinary pressures and temperatures in the presence of titanium chloride-triethylaluminum (Ziegler catalyst) but he did not include propylene in his patent application. Hogan and Banks at Phillips and Zletz at Standard Oil of Indiana polymerized propylene in the presence of metal oxides. Ed Vandenburg at Hercules, S.B. Lippincott of Standard Oil of New Jersey, and B. Wright of Petrochemicals LTD (Shell), like Natta, received licenses to use the Ziegler catalyst.

Bernard Wright of Petrochemicals LTD in England made solid polypropylene on a large commercial scale one day when the HDPE plant ran out of ethylene. Wright and his supervisor T. Barrows did not inform Ziegler of this production since it was assumed that the production of polypropylene was covered by the Ziegler patent. This was untrue. Actually, H. Breil had produced polypropylene at the Max Planck Institut before the Wright production but did not recognize the product as polypropylene.

Breil became impatient when the drop in pressure, during the polymerization of the less volatile propylene was not as great as that with the more volatile ethylene and hence, he added ethylene and observed the anticipated drop in pressure. Later, Ziegler claimed that the product was a copolymer but some polypropylene must have been produced initially before the addition of ethylene.

The polymerization of propylene was repeated successfully by H. Martin some months later in May of 1954. Ziegler then filed a patent application and after sending a sample to Dr. Natta, learned that Natta had filed a patent application on the polymerization of propylene several days earlier.

Dr. Paoli Chini used Ziegler's catalyst to produce polypropylene in Natta's laboratory on March 11, 1954. After seeing the solid product, Dr. Natta wrote in his notebook, "today we made polypropylene."

Dr. Natta attempted to publish an account of this successful experiment in a communication to the editor of The Journal of the American Chemical Society but the paper was rejected by the editor because of lack of experimental data. however, Dr. Paul Flory, who was a member of the editorial board, asked that this decision be reversed and the communication was published in the spring of 1955.

When Ziegler's patent attorney suggested that he include propylene as well as ethylene in his low pressure polymerization patent application, his reply was "Es Geht Nicht." Breil supported this negative attitude when he noted very little decrease in pressure during the formation of polypropylene. When asked why he did not inform Dr. Ziegler that he had produced solid polypropylene, Dr. Natta replied," I did not tell him because I had to take out patents first. I asked him whether he had polymerized propylene in order to know whether my process was new."

Because of familiarity with x-ray crystallography and IR spectroscopy, Natta was able to show that solid polypropylene was a crystalline stereoregular polymer with a three fold helical conformation. As stated in Flory's book, which was read by Natta, C. Schildknecht had previously shown that polyvinyl alkyl ethers, produced at low temperature, were stereospecific polymers. Of course, solid polypropylene had not been reported at the time that Schildknecht conducted his investigations.

Natta used the optically active isomeric terminology DD, DL, etc. when he first described his stereoregular polypropylene. However, his wife Rosita Beati Natta had a better idea and coined the term isotactic "DD", syndiotactic (DL) and atactic (DLLDLDD) to designate these stereospecific polymers. Later, one pundit called her the real creator of isotactic polypropylene.

Natta's group extended the use of the Ziegler catalyst to produce stereoregular polymers from 1-butene, 4-methyl-1-pentene and styrene. Fred Forster of Firestone and Sam Horne of Goodrich used the lithium and Ziegler catalyst respectively, to polymerize isoprene to produce cis-1, 4-polyisoprene which was identical to natural rubber. Later (in 1973), the editors of Chemical and Engineering News, who had a "blimp complex," credited this discovery to Goodyear. Later, another commercial elastomer was produced by the copolymerization of ethylene and propylene in the presence of a diene (EPDM).

Giulio Natta was born at Imperia near Genoa, Italy on February 26, 1903. His father was a judge and it was expected that he, like other members of his family, would study law. However as the result of reading a chemistry book, at the age of 12, he decided to study science.

After graduating from Christopher Columbus High School he majored in mathematics at the University of Genoa and later enrolled in Milan Polytechnic where he was awarded a doctoral degree in chemical engineering in 1924. He received a Libero Docente degree from Milan in 1927 where he served as an assistant professor until 1933.

He spent the next two years as a full professor at the University of Pavia and left to accept a position as professor of physical chemistry at the University of Rome. In 1938, after a year as head of the Institut of Industrial Chemistry at Turin, he returned to Milan and spent the rest of his professional life as a consultant for Montecatini Chemical Company and much of his success in the discovery of isotactic polypropylene may be attributed to continued financial support from that large chemical firm.

Both Giulio and Rosita Natta were fond of the outdoors. She was an avid fisherwoman and he was a collector of fossils and mushrooms. His trips, however, were limited. His mother was killed in an automobile accident and he, not only did not drive, but was even reluctant to ride in an automobile. Nevertheless, he continued as an active director of the Institut but in later life was stricken with Parkinson's Disease and was seriously handicapped in his speech and walk. He was supported by his son Giuseppe when he received the Nobel Prize in 1963. His speech was read by a colleague. He received the Lomonosov Gold Medal of the USSR Academy of Science in 1909 and was awarded honor degrees from many universities.

After his wife's death in 1968 he lived with his daughter Frances and retired in 1973. He died in Bergamo, Italy on May 2, 1979.

In 1948, Dr. Natta concluded that "a revolution will be marked by the development of processes that lead to the formation of macromolecules having a predetermined structure. They will make some branches of industry independent of agriculture and increase the area of land used for the production of food." His contributions to polypropylene plastics, film and fibers as well as ethylene-propylene copolymers and polyisoprene elastomers have helped make his 36 year old prediction come true. Over 6 million pounds of these polymers of propylene were produced in the US in 1987. The largest producer of polypropylene is Himont which is owned jointly by Montedison (successor to Natta's Montecatini) and Hercules where Ed Vandenburg produced polypropylene in the mid 1950's.

REFERENCES

P.J.T. Morris, "Polymer Pioneers," The Center for History of Chemistry, Philadelphia, PA, 1986.

CHAPTER 22

OTTO BAYER
FATHER OF POLYURETHANES

ABSTRACT

Emil Fischer and most students in qualitative organic analysis were familiar with the reactions of monofunctional isocyanates with alcohols, amines and aldehydes. Otto Bayer, who was manager of research of the I.G. Farbenindustrie Laboratories at Leverkusen extended these simple reactions to difunctional reactants and obtained a versatile family of polyurethane fibers, foams, elastomers, plastics and coatings.

Since he recognized the existence of macromolecules, as championed by Hermann Staudinger and was able to produce more wear resistant elastomers than the vulcanized rubber produced by Charles Goodyear, it was appropriate that he be awarded the Hermann Staudinger prize (1973) and the Charles Goodyear medal (1975). He also received many other prestigious awards and was awarded honorary doctoral degrees by six universities in recognition of his unique efforts as a pioneer in polymer science.

INTRODUCTION

In his introduction of Dr. Otto Bayer at the Goodyear Medalist Ceremonies at Cleveland, Ohio, in 1975. Dr. Herman Mark quoted from "The Role of Coincidence in Organic Chemistry" by the 1975 Goodyear Medalist as follows: "A great discoverer will always be a person, in which intuition, the ability to observe, childish curiosity and playfulness, talent for combination and fantasy, tenacity and optimism, expert knowledge and intelligence

R. B. Seymour (ed.), Pioneers in Polymer Science, 213–219.
© 1989 by Kluwer Academic Publishers.

are happily united." Somewhere else, Dr. Bayer wrote, "The most powerful incentives in organic chemistry are still experiment and fantasy. Therefore, the intuitive artistic principle and not the deductive-mathematical principle will lead us to new shores."

When expanded and related to specific discoveries, these statements are inordinately descriptive of this great twentieth century chemist. This great discoverer was primarily an organic chemist. Yet, he was the prime mover in launching new editions of Gmelin's "Handbook of Inorganic Chemistry."

Otto Bayer was intuitive, observant, and curious. Thus, through his expertise and intelligence, he was able to lead his company (I.G. Farben-industrie and its successor, Bayer AG) to new shores in a wide variety of fields.

He directed the development of new dyestuffs and optical brighteners, but as a research manager, recognized that these improved products would merely replace existing products manufactured by his employer. He utilized intuitive artistic principles to develop the first effective stabilizing drug for tuberculosis and for the synthesis of new phosphorus insecticides.

As manager of the Central Research laboratory of I.G. Farben-industrie at Leverkusen in the 1920's, he directed the work of Tschunker, Bock, Konrad, and Kleiner, which led to Buna S and Buna N which are the world's most widely used synthetic elastomers. This dyestuff's chemist accepted the concepts promulgated by Hermann Staudinger and Wallace Carothers and became one of the world's best known polymer chemists.

Chemists at Leverkusen were familiar with cellulose ester plastics and poly 2,3-dimethylbutadiene rubber, but these were produced on a small scale without the benefit of a knowledge of macromolecular chemistry. The production of Buna S, which was a copolymer of butadiene and styrene, which was aa laboratory curiosity. In spite of opposition from influential members of the board of directors of I.G. Farben-industrie, Bayer directed the development of a process for the production of styrene monomer and polystyrene at the Ludwigshafen and Bitterfield plants.

The production problems with Buna N, which is a copolymer of butadiene and acrylonitrile, required even more tenacity and optimism, but Dr. Bayer was successful in the development of a commercial process for the catalytic hydrocyanation of acetylene for the production of acrylonitrile. This process which was used worldwide for many years has been displaced by the catalytic ammoxidation of ethylene process.

Buna S, under the name of GRS, and phosphorus insecticides are the most widely used elastomers and insecticides today, but Otto Bayer's greatest achievements were in the field of polyurethanes in which he was able to utilize his playfulness, expert knowledge and intelligence, and most of the other attributes that he used to describe the great discoverer.

This great discoverer was born at Frankfurt on the Main in 1902, and after studying under Julius von Braun, received his Ph.D. degree from the University of Frankfurt at the age of 22. his first and only employment was with I.G. Farben-industrie where he rose rapidly from a research chemist at Casella to Director of Research at Bayer AG in 1951.

He, like other organic chemist, was aware of the reactions of organic isocyanates with alcohols and amines for the quantitative, exothermic, room temperature production of urethanes and ureas, respectively. Actually, Bayer had characterized a large number of new amines by the production of ureas from the action of isocyanates.

Accordingly, it is not surprising to note that he proposed the use of organic diisocyanates and diols or diamines for the production of macromolecules in 1936. However, these suggestions were not accepted by his superiors who now questioned his ability to remain as head of an industrial research laboratory. Even one of his colleagues told him politely that "if you had ever made a monoisocyanate, you would never have had that crazy idea."

However, Dr. Rinke was successful in his attempt to synthesize octane-1,8-diisocyanate from sebacic acid and sodium azide. In demonstrating his childlike curiosity, Dr. Bayer added the first few drops of the distillate to 1,4-butane diol and obtained a product which could be drawn into filaments, much like Carothers' nylon.

Subsequently, Bayer and coworkers, obtained hexane diisocyanate from the reaction of 1,6-hexamethylenediamine hydrochloride and phosgene. This product was reacted with butane diol to produce several kilograms of aliphatic polyurethane. The melting point of this polymer (184°C) was too low for textile fiber applications, but it is still used for the production of bristles.

In spite of Bayer's tenacity and optimism, the patent attorneys at I.G. Farben-industrie refused to file a patent application since they believed that the publication in a journal would suffice to preserve priority for this laboratory curiosity.

Of course, there were other almost insurmountable hurdles which would hinder the production of polyurethanes. After some time, procedures were developed for the production of phosgene on a larger scale, and hydroxyl terminated polyesters were used in place of butane diol in the production of polyurethanes.

The latter may be attributable to the intuitive-artistic principle because smaller quantities of the expensive diisocyanate were required for the production of polyurethanes from these hydroxylterminated polyesters. In addition, Bayer and coworkers were able to synthesize a less expensive diisocyanate from the reaction of phosgene and 2,4-toluenediamine. This tolulyl diisocyanate (TDI) is one of the most widely used diisocyanates today.

Bayer and his coworkers discovered that urethanes based on phenols decompose to isocyanates and phenol at 130°C. This discovery led to the use of these blocked isocyanates for use in baking enamels.

Subsequently, these investigators obtained nonvolatile polyisocyanates by the reaction of glycols with an excess of the diisocyanate. These investigations led to the production of "fast drying" urethane modified linseed oils.

In spite of these successful experiments, there was little if any commercial use for polyurethanes. According to Dr. Bayer, the "great breakthrough" in polyurethane chemistry occurred in 1941 when Drs. Hoechtlen, Droste, and Bunge made a porous casting which those in the testing laboratory named an "imitation swiss cheese." The bubbles were the result of the production of CO_2 from the reaction of isocyanate groups with unreacted carboxyl groups in the polyester. Subsequently, Bayer used small amounts of water to generate carbon dioxide. Almost 2.7 billion pounds of these polyurethane foams were produced in the US in 1987.

However, the transition from "imitation swiss cheese" to one of the most important cellular materials required tenacity, optimism, and expert knowledge. Organic tin compounds developed by Presswork were used to catalyze the reaction of diisocyanates with glycols, triethylenediamine (Dabco), developed by Houdry, was used to catalyze the reaction of diisocyanates with water and silicon surfactants were used to produce uniform cells.

After much research, elastomers with a high degree of elasticity, excellent abrasion resistance, and high tear strength were produced from the reaction of 1,5-naphthalene diisocyanate. The outstanding adhesive properties of these polyurethanes was observed when some of these elastomers stuck tenaciously to the metal mold surface.

Bayer's tenacity and diplomacy were demonstrated when he was able to continue his experimentation in spite of administrative regulations and occasional air raids during World War II.

After a low productive decade, the Mobay Corp. was formed by Bayer and Monsanto for the production of polyurethanes in the U.S. in 1953. However, the first polyester-based polyurethane foam disintegrated in humid air and could not compete with rubber latex foam. This difficulty was overcome by the use of hydroxyl terminated polyethers instead of polyesters.

In addition to receiving the Goodyear Medal (1975), Dr. Bayer also received the Adolph von Baeyer Medal

(1951), the Gauss-Weber Medal (1952), the Siemens Ring (1960), the Carl Duisberg Medal (1953), the Otto N. Watt Medal (1966), and the Harries Medal.   He was named Honorary Professor of the University of Cologne and played a major role in launching the new edition of the multivolumed "Hougen-Weyl, Methods in Organic Chemistry."   Dr. Bayer received honorary degrees from universities at Bonn, Cologne, Mainz, Munich, and Aachen. He was also recognized throughout the Bayer Company as an excellent manager.

Otto Bayer who has been called the "father of polyurethane chemistry" died on August 1, 1982, a few months before his 80th birthday.

REFERENCES

O. Bayer, Rubber Chemistry and Technology, 48, 73 (1975).

G. Gerlel, "Polyurethane Handbook," Hanser Publishers, Munich, Germany, 1985.

R.B. Seymour, in Encyclopedia of Physical Science and Technology, Vol. XI, Academic Press, Orlando, FL, 1987.

G. Woods, "The ICI Polyurethanes Book," John Wiley, New York, NY, 1987.

CHAPTER 23

JOHN P. HOGAN AND ROBERT BANKS
INVENTORS OF LINEAR POLYOLEFINS

Since honors, such as the Nobel Prize, are not dependent on scientific evaluations and are seldom awarded to industrial scientists, it is not surprising that the true inventors of linear polyolefins were not recognized as nobel laureates. Of course, the early investigations of high density polyethylene (HDPE), crystalline polypropylene (itPP) and linear low density polyethylene (LLDPE) occurred in several different laboratories by different skilled researchers who were aware of previous developments in this field. Hence, it is not surprising that neither "Speed" Marvel, Paul Hogan, Bob Banks, Ed Vandenburg, Alex Zletz, Don Carmody, or Grant Bailey were recognized as the inventors of HDPE.

Of course, the polyethylene produced by the decomposition of diazomethane by von Pechmann, Bamberger and Tschirner in the 1890's, by Meerwein and Burnelit in 1928, by Weile in 1938 and by Buckley, Cross and Ray in 1950 was linear polyethylene (HDPE). Likewise, the polyethylene produced by "Speed" Marvel in 1930 and by Frank Mayo in the 1940's, who used coordination catalyst systems was HDPE but there was little interest in this polymer at that time. Actually, some chemists, including nobel laureate Hermann Staudinger maintained that it was not possible to polymerize ethylene.

Nevertheless, in addition to syntheses by Marvel and Mayo, one of my classmates, Dr. Grant Baily working with L.S. Reid produced low molecular weight polyethylene by passing ethylene over a nickel oxide/silica-alumina catalyst. These investigators also made solid polypropylene from propylene via the same route in 1945. Chemists at Shell

R. B. Seymour (ed.), Pioneers in Polymer Science, 221–225.
© *1989 by Kluwer Academic Publishers.*

obtained comparable solid polymers but since they were seeking high octane gasoline, they disregarded these syntheses which plugged up their equipment.

Had these investigators been polymer scientists, they would have known about the success of the investigations of low density polyethylene (LDPE) which led to the commercial production of this branched polymer in the late 1930's. However, their objective was to convert gaseous ethylene and propylene into high octane gasoline.

Fortunately, for Phillips Petroleum Co., John Paul Hogan and Robert L. Banks followed up the Baily-Reid investigations and applied for patents on the synthesis of linear polyolefins including polyethylene, polypropylene and polymethylpentene. In 1951, A. Zletz of Standard Oil of Indiana, working with another classmate of mine, Don Carmody, also patented a low pressure process for making HDPE.

As discussed in a separate chapters, Karl Ziegler patented a low pressure process for the production of HDPE in 1955 and G. Natta patented a low pressure process for the production of crystalline polypropylene in 1954. However, this patent was declared invalid and the patent was granted to Hogan and Banks in 1983.

The preceding discussion might be dismissed as a minor controversy concocted by money hungry patent lawyers until it is recognized that the U.S. production of these polyolefins in 1987 was as follows:

LDPE	9.5 billion pounds
HDPE	7.8 billion pounds
PP	6.4 billion pounds

The data given for LDPE also include 3.5 billions pounds of linear low density polyethylene (LLDPE). LLDPE, which was also produced by Hogan and Banks in the 1960's is a copolymer of ethylene and higher alkenes, such as 1-butene.

Both K. Zeigler and G. Natta had been awarded Ph.D. degrees and had close university affiliations. However, Hogan and Banks possessed only M.S. degrees and were employed by Phillips Petroleum Co. at Bartlesville, OK which is a long way from any university and is a gasoline oriented corporation.

John Paul Hogan was born in Lowes, KY, on August 7, 1919. After receiving a B.S. degree from Murray State University in 1942, he taught science courses at Mayfield, KY, High School in 1942-43 and taught a special course in physics for U.S. servicemen at Oklahoma State University (1943-44). While teaching at OSU, he earned enough credits to obtain an M.S. degree. Murray State also gave him an honorary D.Sc. degree in 1971.

Dr. Hogan is a recipient of 40 patents from the U.S. Patent office. He received the Creative Inventor Award from the American Chemical Society in 1959, was named inventor of the year by the Oklahoma Bar Association in 1976, was named a chemical pioneer by the American Institute of Chemists in 1972, and a pioneer in polymer science by Polymer News in 1981. He and Robert Banks were corecipients of the Perkin Medal in 1986.

John married Glenda Maultrie in 1943. The Hogans are parents of Fay, Kenneth and Susan Hogan. Since there are only 51 former vice presidents of the U.S., Lowes is proud to be recognized as the birthplace of Alben Barclay. Yet there are only two coinventors of linear polyolefins and one of these was also born in Lowes, KY.

In addition to sharing the Perkin Medal with coinventor J.P. Hogan, Robert L. Banks, was the recipient of the Petroleum Chemistry award by the American Chemical Society in 1978, the Oklahoma Chemistry award in 1974 and the Chemistry Pioneer award by the American Institute of Chemists in 1981. He was named a Pioneer in Polymer Science by Polymer News in 1985.

Like his coinventor, Bob was born in a small town viz Piedmont, MO on November 24, 1921. He was the youngest of a family of seven children parented by Dentist Dr. James A. and music teacher Maude McAlister Banks. After graduating from Piedmont High School in 1940, Bob enrolled in Southwest Missouri State College, but transferred after his sophomore year to the University of Missouri at Rolla where he received the B.S. degree in chemical engineering in 1944. He was awarded an M.S. degree by Oklahoma State University in 1953 and was the recipient of an honorary professional degree and an alumni merit award from Rolla in 1976 and 1980.

His first position was as a process engineer for Co-op Refinery at Coffeyville, KS but he left in 1945 to accept a position with Phillips Petroleum Co. at Bartlesville, OK. Robert Banks was awarded 48 patents by the U.S. Patent office and is the author of 35 publications in technical journals. He is a member of the American Chemical Society, the Catalyst Society and a fellow of the American Institute of Chemists.

Bob married Mildred K. Lambeth in 1947. The Banks are parents of Susan Lee, Mary Kathleen and Melissa Ann. They also have four grandchildren. Since retiring, Bob has been playing golf at Hillcrest and Sunset Country Clubs in Bartlesville. He is proud to be using tees molded from HDPE and golf balls covered with an ionomer copolymer of ethylene and methacrylic acid.

REFERENCES

Seymour, R.B., Cheng, T. eds. "History of Polyolefins," D. Reidel Publishing, Co., Dordrecht, The Netherlands, 1986.

Fredrich, M.E.P., Marvel, C.S, J. Am. Chem. Soc., 52, 376 (1930).

McMillan, F.M., "The Chain Straighteners," McMillan Press, London, 1979.

Seymour, R.B., ed., "History of Polymer Science and Technology," Marcel Dekker, New York, NY, 1982.

Seymour, R.B., The World and I, 3, (2), 188 (1988).

Fawcett, E.W., Gibson, R.O., et al, Brit. Pat. 47,590 (1937) (LDPE).

Hogan, J.P., Banks, R.L., U.S. Pat. application 333,576 (1953) (HDPE).

Zletz, A., U.S. Pat. 2,692,257 (1957).

Ziegler, K., Breil, H., et al, U.S. Pat. 3,257, 332 (1966).

Natta, G., et al, U.S. Pat. 3,715,344 (1977).

Hogan, J.P., Banks, R.L., U.S. Pat. 4,376,851 (1983).

Seymour, R.B., Fisher, C.H., "Profiles of Eminent American Chemists," Litarvan Enterprises, Sydney, Australia, 1988.

CHAPTER 24

MODERN PIONEERS IN PLASTICS, FIBERS, INORGANIC POLYMERS, ELASTOMERS AND ENGINEERING POLYMERS

HERMAN ALEXANDER BRUSON

While some polymer chemists are better known for their accomplishments, few excelled Herman Alexander Bruson in ingenuity and productivity. Herman, who was the son of Samuel J. and Rebecca Arnowitz Bruson, was born in Middleton, OH on July 20, 1901. After receiving the B.S. degree from MIT in 1923, he enrolled in Zurich's Federal Technical University where he was awarded the Ph.D. degree in 1925.

After spending 3 years (1925-1928) as a research chemist at Goodyear Tire and Rubber Company, he joined the research group at Rohm and Haas where he remained for 20 years. When fiber producing firms learned of the successful production of acrylic fibers by DuPont and Monsanto (Chem Strand), they were determined to try to enter the synthetic fiber business. Hence, it was no surprise when Industrial Rayon Corp. offered Dr. Bruson a lucrative position as a research manager in 1948.

Since he was unable to obtain adequate funding for a new plant facility, he left Industrial Rayon in 1952 to join Olin Corp. where he served as vice president until his retirement in 1966. The National Association of Manufacturers established a pioneer award to recognize those few inventors who were awarded 40 or more patents by the U.S. Patent Office. This award was discontinued after being given to Dr. Bruson in recognition of his being granted more than 500 patents.

R. B. Seymour (ed.), Pioneers in Polymer Science, 227–250.
© 1989 by Kluwer Academic Publishers.

Herman was one of the first to demonstrate the versatility of polymers by developing application for acrylic polymers as oil additives, antioxidants, plasticizers, fibers, coatings and surfacants, as well as related compounds for flame retardants, foams, bactericides and pesticides.

Herman was a member of the American Chemical Society, The American Association for the Advancement of Science, the American Institute of Chemists and the Chemists Club. I always looked forward to informal discussions with Herman at the Gibson Island Conferences in the 1930's and 1940's. He was the recipient of the Wisdom Hall of Fame award, the American Chemical Society's Creative Invention award, the Eli Whitney award and the Maurice R. Chamberland award.

Herman married Virginia Haber in 1929. The Bruson's are parents of Rita, Dorothy and Barbara. Dr. Bruson died in Woodbridge, CT in 1981.

ARTHUR MAYNARD BUECHE

Few sons could be given better advice than that given by the mother of Arthur Maynard Bueche. She discouraged his goal of being a lawyer by stating that being a chemist was a more honorable way of making a living.

Arthur was born in Flint, MI on November 14, 1920. After graduating from Flint Junior College and the University of Michigan (1943) where he demonstrated his ability as a scholar, athlete and poet, he enrolled in graduate school and was awarded the Ph.D. degree from Cornell University in 1947. After spending three years on the faculty at Cornell, he joined GE where prior to his death in 1981, he demonstrated unusual ability as a polymer scientist and research manager.

Dr. Bueche was a member of the American Chemical Society, the American Institute of Chemists, the American Institute of Physics and the American Association for the Advancement of Science. He was elected to membership in the National Academy of Sciences and the National Academy of Engineering and was awarded Medals by the American Society of Metals (1978), the Industrial Research Institute (1979) and the American Society for Mechanical Engineers (1980). He was the father of two sons.

PETER JOSEPH WILLIAM DEBYE

The unusual scientific ability of Peter J.W. Debye was recognized in 1911 when he was appointed as the successor to Albert Einstein at the University of Zurich. In spite of the prestige of this position, he left to accept a position as professor of theoretical physics at Utrecht in 1912 but returned to Zurich as professor of experimental physics in 1920.

This world class scientist who was awarded the Nobel Prize in Chemistry in 1936, was born in Maastricht, Holland on March 24, 1884. He was the son of Marie Reumkens Dibje and Wilhelmus Debye. After graduating from Hoogre Burer School at Maastricht in 1901, he enrolled in the Techniche Hochschule at Aachen where, while still a student, he published his first work on Foucoult Currents in 1907.

He was awarded the Ph.D. degree in 1908 while serving as a professor of theoretical physics at the Ludwig-Maximillian University in Munich. While at the University of Zurich (1920-1927) he collaborated with Dr. Huckel in the development of the Debye-Huckel theory of electrolytes in solution.

After joining the faculty at the University of Leipzig in 1927, he continued his investigations of dipoles and electrolytes and measured interatomic distances by x-ray scattering techniques. He continued these investigations at the University of Berlin where he was appointed a professor and director of the Max Planck Institute in 1935.

He emigrated to the U.S. in 1939 when he accepted the Baker lectureship at Cornell where he remained as chairman of the chemistry department until his retirement in 1952. He left Berlin since he refused to apply for German citizenship but he became an American citizen in 1946.

While much of his reputation was based on nonpolymeric accomplishments, such as demonstrated by the Debye-Huckel theory, the Debye-Scherrer x-ray diffraction technique, the Debye-Sears effect in liquids, the Debye temperature, the Debye shielding distance, the Debye frequency and the Debye unit of electric moment, his development of the light scattering technique for the determination of the molecular weight of polymers resulted in his also being recognized as a world class polymer scientist.

He was a frequent delegate in the 1930's and 1940's to the Gibson Island Conferences where he was most generous in discussing his polymer science investigations at Cornell.

In addition to having an award named after him by the American Chemical Society, he was awarded the Priestley Medal and was given honorary degrees by several universities. The city of Maastricht recognized this local scientific genius by erecting a statue of Debye in its city hall.

JAMES ECONOMY

One of the pioneer moldable high temperature polymers was developed by Dr. James Economy while employed in the laboratories of Carborundum (1960-1975). His melt spinnable aromatic copolyester fibers (Ekonol) paved the way for the development of liquid crystal polymers by Dartco in 1984.

James was born in Detroit, MI on March 28, 1920. He was the son of Bersi Lalouise and Peter George Economy. After receiving his B.S. and Ph.D. degrees from Wayne State University (1950) and the University of Maryland (1954), he worked for two years as a postdoctoral fellow under the direction of Dr. Marvel at Illinois. He then spent four years (1956-1960) at Allied Chemical Co. before accepting a position as general research leader at Carborundum. He has served as manager of polymer science and technology at IBM at San Jose, CA since 1975.

Dr. Economy is a member of the National Academy of Sciences, the American Chemical Society, the International Union of Pure and Applied Chemistry and the American Institute of Chemists. He has been awarded more than 70 patents by the U.S. Patent Office and is the recipient of fourteen IR-100 awards from Industrial Research. He is the recipient of the Schoelkopf Medal (1972), the Southern Burn Institute award (1976), the American Chemical Society Phillips Polymer Science award (1985) and the American Institute of Chemists Chemical Pioneer award (1987).

James married Stacy Zapantis in 1961. The Economy's are parents of Elizabeth, Peter, Katherine and Melissa.

CARLTON ELLIS

Carlton Ellis, who was one of the most prolific inventors of all times (753 U.S. patents), was born in Keene, NH on September 20, 1876. He graduated from MIT in 1900 and served as an instructor at his alma mater for two years.

In 1905, he formed the Chadeloid Co. which produced some of his patented products, including a paint remover which he formulated before graduating from MIT. His ability to invent and patent became so well known that few firms would permit him to view their production facilities.

In cooperation with Nathaniel Foster, he formed the Ellis-Foster Laboratories in Monclair, NJ and Key West, FL. This firm patented over 25 inventions annually for a period of 25 years.

These inventions were in many fields from food products to cracked petroleum but his preference was polymer technology. He was awarded patents by the U.S. Patent Office on the first automobile enamel, urea formaldehyde plastics and fiberglass-reinforced polyester plastics.

Carlton was a prolific author. The best known of his nine books was <u>The Chemistry of Synthetic Resin</u> which was published by Reinhold in 1935. Every major development in plastics prior to the 1930's was discussed in this book. An attempt to update this reference in the early 1940's resulted in a nervous breakdown by the new author. Few had the ability to review the entire plastics industry and the developments in the decade following the publication of the Ellis book exceeded all the developments prior to the 1930's. Carlton Ellis died in Miami Beach, FL on January 13, 1941.

DANIEL WAYNE FOX

Polycarbonates, which are recognized as premier tough heat resistant, transparent plastics were not available commercially prior to the mid 1950's, until Dr. Herman Schell of Farbenfabriken Bayer AG and Dr. Daniel Wayne Fox of GE synthesized these polymers independently. Because of the unwillingness of Montedison to recognize that Hogan and Banks of Phillips Petroleum Co. were the true inventors of crystalline polypropylene, many lawyers became millionaires during the 20 years of litigation on the polypropylene patent. Since Farbenfabriken Bayer and GE preferred to support polymer scientists rather than lawyers, these firms agreed to cross-license these important engineering polymers which are marketed as Merlon and Lexan.

Daniel, who was the son of Marie Alma Hill and Daniel Francis Fox, was born in Johnstown, PA on May 14, 1923. After serving as Lieutenant in the U.S. Air Force (1942-1945) he enrolled in Lebanon Valley College where he was awarded the B.S. degree in 1948. He was awarded the M.S. and Ph.D. degrees by the University of Oklahoma in 1951 and 1952.

He joined GE in 1952 and has continued to be a productive researcher and administrator at GE since that time. Dan is a member of the National Academy of Engineers, the American Chemical Society, Society of Plastics Engineers, Phi Lambda Upsilon and Sigma Xi. He has been granted more than 50 patents by the U.S. Patent office and has authored several books including Aromatic Polycarbonates.

Dr. Fox was inducted into the Plastics Hall of Fame (1976), received the International award from Society of Plastics Engineers in 1985, was named a Pioneer in Polymer Science by <u>Polymer News</u>, a Chemical Pioneer by the American Institutes of Chemists in 1987 and received the Midgely Medal in 1988.

Dan married Martha Joyce Schmidt in 1948. The Foxes are parents of Barbara, Ann and Daniel S. Fox. Dr. Fox died February, 1989.

CHARLES GOODYEAR

Charles Goodyear was an uneducated inventor and not a polymer technologist. Nevertheless, his serendipitous discovery of a process for curing (cross-linking) rubber, which is still in use today, has had a profound influence on polymer science and technology. If not engraved in stone, his name was molded in rubber when the Seiberling Brothers used his name for their rubber plant in Akron, OH.

Charles, the son of Amasa Goodyear, an inventor and hardware manufacturer, was born in New Haven, CT on December 29, 1800. After a minimal education he dropped out of Naugatuck High School in 1817 and then served as an apprentice to another hardware manufacturer in Philadelphia. He joined his father in the hardware business in 1821, and they opened the nation's first retail hardware store in the US in 1826. Unfortunately, because of Goodyear's liberal credit policies, this business was bankrupt in 1830, and Charles then became a full time inventor of mechanical devices.

Charles, like many other inventors of that time, was familiar with the effect of temperature on waterproof textiles, such as the Macintosh rain coat and hoped to solve this problem by a series of empirical experiments. In spite of his indebtedness, he did acquire The Eagle India Rubber Co. in Woburn, MA, which then acquired Nathaniel Hayward's patents on a mixture of natural rubber, which he called gum elastic, and sulfur, which could be cured in the sunlight, which was not always available on a daily basis in New England.

The serendipitous discovery in 1839 that a mixture of sulfur and natural rubber could be crosslinked (vulcanized) when dropped accidently on his wife's hot kitchen stove was the breakthrough that made the use of rubber practical for applications other than erasers.

This simple process was patented by Goodyear in 1844, but about sixty other inventors used Goodyear's vulcanization process.   Combating these cases cost Goodyear time, which he could ill afford, and money, which he didn't have. However, Daniel Webster succeeded in winning the Great India Rubber case in 1852.

His patent was denied in England and invalidated in France. Because of combating infringement cases, Charles continued to spend more time in debtors' prison than in his plant.   Nevertheless, he was made an officer in the Legion of Honor in France, and he received several other awards.

In spite of his empiricism, he authored a two-volume book on the discovery of the vulcanization of Gum-Elastic and its applications.

Charles married Clarissa Beecher in 1824.   The Goodyear's were parents of nine children. After the death of Clarissa in 1853, he married Fanny Wardell and three children were born from this union.   Many of Charles ancestors were inventors. Hence, it was not surprising that his brother, Nelson Goodyear, invented hard rubber (ebonite) and his son Charles, II invented the welt machine and established the Goodyear Welt Shoe Machine Company.

JOHN GOUGH

In 1802, almost four decades before the discovery of the process of curing (vulcanization) natural rubber by Charles Goodyear, a blind scientist used his other senses such as his lips, to detect the increase in temperature when a band of india rubber (caoutchouc) was stretched. This increase in temperature could not be detected by the crude thermometers that were available in the early 19th century but was readily detected by Gough's sensitive lips.

John Gough, who at the age of three lost his sight after an attack of small pox, was born in the English Lake District in 1757. He was the son of a shearer-dyer who helped John to overcome his disability by teaching him to develop his touch and tongue in order to visualize his surroundings. He was also encouraged to conduct experiments in his father's dye house.

John's stretching vs temperature experiments were shown to be reversible i.e., a stretched rubber band contracted when heated and elongated when cooled. When thermometers which were as sensitive as John Gough's lips became available in 1859, Joule quantified Gough's experiments. John also showed that a stretched rubber band lost much of its elasticity when placed in cold water but this property was recovered when the water was heated.

John demonstrated that stretched rubber, which had a melting point of 26°C, crystallized at about 0°C. In addition to the quantification of his experiments by Joule, these fundamentals became the basis for the modern theory of elasticity which were further refined by nobel laureate Paul Flory in the 1930's.

While employed as a tutor for John Dalton, William Whewell and Joshua King in Cambridge, John invented a form of abacus which was a forerunner of the modern computer. In addition to being recognized as a world class mathematician, he was also the author of 50 reports in philosophical journals.

In the Excursions, Wadsworth wrote the following about John Gough:

Me thinks I see him - how his eyeballs rolled
Beneath his ample brow in darkness pared
But each instinct with spirit and the frame
Of the whole Countenance alive with thought.

ERNST ALFRED HAUSER

Rubber molecules were erroneously considered to be colloids by many scientists in the 19th century. These molecules are, of course, macromolecules but these molecules are present as clusters of many molecules i.e., are in the form of colloids in natural rubber latex.

In 1924, Ernst Alfred Hauser used an opal red dye to make these colloidal particles more visible when viewed under a microscope.

In collaboration with Frendlich in 1925, Hauser proposed that the stability of rubber latex was due to an adsorbed protein outer layer. In 1926, he collaborated with Herman Mark on x-ray studies of stretched rubber and assigned unit cell parameters which are approximately the same as those accepted today.

Ernst was born in Vienna on July 20, 1896. He joined the Austrian Army in 1914 and after his discharge in 1918, enrolled in the University of Vienna where he was awarded the Ph.D. degree in 1922. Prior to serving as a visiting associate professor at MIT (1928-30), he held several research positions at Gottingen and Frankfurt where he was associated with Max Born and Richard Zsigmondy.

He emigrated to the U.S. in 1935 and was promoted to a full professor at MIT in 1937. In addition to serving as director of colloid chemistry at MIT, he also served as professor of colloid chemistry at Worcester Polytechnic Institute (1948-52) where he was awarded an honorary D.Sc. degree in 1952.

As a result of his many publications on the colloidal properties of rubber latex, including Latex (1947), Colloidal Chemistry of Rubber (1928), Colloidal Phenomena (1939), Experiments on Colloid Chemistry (1940) and Silicia Science (1955) and his editorship of Handbuch der Gesamten Kautschuktechnologie, he became recognized as ont of the world's leading latex chemists.

Ernst and Vera M. Fischer, whom he married in 1902, were parents of Ernst, Wolf and George. The senior Ernst Hauser died at Cambridge, MA on February 10, 1956 He was buried in New Hampshire at Cathedral of the Pines.

ROELOF HOUWINK

While few modern scientists will deny the direct relationship of molecular weight of polymers to their viscosity, this was not always true. In the 1920's, both Staudinger and Mark proposed such a relationship and in 1929 Staudinger derived his famous "viscosity law" based on the Einstein relationship of viscosity and concentration of solute spheres. Staudinger assumed that the macromolecules were in the form of rigid rods which rotated in one direction only. Because of these compensating errors, Staudinger's viscosity law, $\eta_{sp}=KM$, in which the specific viscosity was proportional to the molecular weight, was a good approximation of the correct relationship.

In contrast, Mark maintained that macromolecules could assume many different conformations (shapes) and in collaboration with Guth and Kuhn, proposed a power form for the Staudinger equation i.e., $\eta_{sp}=KM^a$. A similar equation was proposed simultaneously by Roelof Houwink and the above equation is now referred to as the Mark-Houwink viscosity equation.

Roelof Houwink was born in 1897 in Meppel, Holland. He majored in botany and received the Ph.D. degree at the University of Delft in 1934 but his studies of the hevea rubber plant catalyzed an interest in rubber which he maintained throughout his career in the Dutch Government Rubber Service and at the Vredesteen rubber factory.

In 1925, while employed by Philips at Einhover, he developed an interest in phenolic resins which was enhanced by a visit to the GE plant at Pittsfield, MA and by discussions with Leo Baekeland. His attempts to conduct major investigations in rubber science at Rubber-Stichting were hampered by World War II but he was able to establish the Plastics and Rubber Research Institute at Delft in 1945. However, interest in natural rubber declined when Indonesia gained its independence in 1956.

In the same manner in which he switched his interests from botany to polymer science, Houwink now switched to atomic energy and was appointed technical director of the

Reactor Centrum in the Hague in 1957. He retired in 1962 but continued his interest in science.

In addition to his fame as one of the developers of the Mark-Houwink equation and as the recipient of many honors, Dr. Houwink was one of the few scientists to publish a prediction in <u>Modern Plastics</u> (1966) that the volume of polymers would surpass that of iron in 1983. His prediction, which was shared by a few other polymer scientists, proved to be accurate and now there is a need to predict the year when the volume of polymers will double that of steel.

STEPHANIE L. KWOLEK

Stephanie L. Kwolek, who is the only women named as a Pioneer in Polymer Science by <u>Polymer News</u> (1987) was born in New Kensington, PA on July 31, 1923. She received the B.S. degree from Carnegie Mellon University in 1946 and was awarded an honorary D.Sc. degree by Worcester Polytechnic Institute in 1981.

She accepted a position with DuPont in 1971 and has continued to investigate aramid fibers since that time. She has been awarded 16 patents by the U.S. Patent Office and is the author of 25 reports in scientific journals. In addition to being a member of the American Chemical Society, Stephanie is also a member of the American Institute of Chemists, Phi Kappi Phi, Sigma Xi and the Franklin Institute.

She is the recipient of the Howard Potts Medal (1976), the ASM International award (1978), the Chemist Pioneer Award from the American Institute of Chemists (1980), the American Chemical Society award for creative inventions (1981), the Society of Plastics Engineers Technology award

(1985) and was inducted into the University of Akrons Polymer Processing Hall of Fame (1985).

In spite of the fact that most minerals, glass and hydraulic cements are polymers, few chemists have investigated inorganic polymers. Some have investigated polyphosphazenes ($-N=P(OAr)_2-$) and silicones ($-Si(R_2)O-$).

R. BRUCE MERRIFIELD

Emil Fischer synthesized polypeptides using techniques available in the early 1900's. Vincent du Vigneaud, who was a student of "Speed" Marvel, upgraded these techniques and produced several biologically active polypeptides, including oxytocin.

Such syntheses were extremely tedious until Bruce Merrifield developed an automated process in which the first member of the peptide chain was anchored to the matrix of a crosslinked polystyrene bead. The propagation was continued using this solid phase synthesis which was described in an article in J. Am. Chem. Soc. in 1963.

R. Bruce Merrifield was born in Ft. Worth, TX in 1921 but since his father was a travelling salesman, he spent his early years in 40 different California schools before graduating from Montebello High School in 1939. He was awarded the B.A. and Ph.D. degrees from UCLA in 1943 and 1949. He was also the recipient of honorary D.Sc. degrees from five other universities.

Bruce is a member of the National Academy of Science, the American Chemical Society, the American Institute of Chemists, Alpha Chi Sigma, Phi Lambdan Upsilon and the American Society of Biological Scientists. He was a recipient of the following awards: Lascher award (1969), Gardner award (1970), Intrascience award

(1970), American Chemical Society Award for creative work in synthetic organic chemistry (1972), Nichols Medal (1973), Instrument special award (1977) and the Alan E. Precia award. He was named a Pioneer in Polymer Science by <u>Polymer News</u> in 1983 and received the Nobel prize in 1984.

Bruce married Elizabeth Findly the day before receiving his Ph.D. degree and the day before accepting a position on the faculty of Rockefeller University where he was appointed full professor in 1966. The Merrifields are parents of six children.

PAUL W. MORGAN

Wallace Carothers shelved his aliphatic polyesters because of their low softening points. However, R.T. Winfield and J.T. Dickson were able to overcome this deficiency by using an aromatic dicarboxylic acid (terephthalic acid) instead of the aliphatic adipic acid.

Polyamide fibers with much higher softening points than nylon 66 or nylon 6 were produced by a team of DuPont chemists who used aromatic reactants instead of the aliphatic reactants used to produce nylon 66. These aramids which were the first liquid crystal polymers, were spun to produce high tenacity fibers which were sold under the trade names of Kevlar and Nomex.

These investigations by Paul Morgan, Stephanie Kwolek and associates stimulated a new era in high modulus temperature-resistant fibers. In contrast to classic strong fibers with tenacities of about 10g/denier, these aramids have tenacities in the order of 30g/denier and moduli greater than 300g/denier. Nomex, which has a melting point of 370°C, is produced by the condensation of terephthalic acid and hexamethylenediame. Kevlar 49, which is resistant to temperatures as high as 300°C is produced by the condensation of p-aminobenzoyl chloride hydrochloride ($HCl.NH_2-C_6H_4-COCl$).

Paul W. Morgan was born in West Chesterfield, NH on August 30, 1911. After graduating from Tomaston (ME) High School, he enrolled in the University of Maine where he was awarded the B.S. degree in 1937. He received the Ph.D. degree from Ohio State University in 1940 and was a postdoctoral fellow at Ohio State where his research advisor was Professor M.W. Wolfrom.

Paul joined the DuPont company in 1941 and retired as a senior research fellow in 1976. He was awarded 37 patents by the U.S. Patent Office and has published about 40 reports in scientific journals and a book on "Condensation Polymers". He is a member of the American Chemical Society, Phi Kappa Phi, Tau Beta Pi, Phi Lambda Upsilon and Sigma Xi. He has served as National Councilor and Chairman of the Polymer Group of the American Chemical Society and chairman of the Delaware Section of the American Chemical Society.

Dr. Morgan is a member of the National Academy of Engineering and the recipient of the Delaware Section Award and the Witco American Chemical Society Award (1976), the Swinburne Medal (1978), the Howard Potts Medal (1976) and the ASM International award (1978). He was named a Pioneer in Polymer Science by Polymer News in 1981.

Paul married Elsie Bridges in 1969. The Morgans are parents of two children.

IWAN IWANOWITCH OSTROMISLENSKY

Iwan Iwanowitch Ostromislensky was one of the few top polymer scientists to enjoy careers both in the USSR and USA. Iwan, who was the son of an officer in the Imperial Guard was born in Moscow on September 8, 1880. His numerous degrees include those from Moscow Military Academy (1895) and Moscow Pilotekhnekum (1899), Ph.D. and M.D. degrees from the University of Zurich (1902 and 1906) and a degree in chemical engineering from Karlsruhe (1907).

After five years on the faculty at Moscow University (1907-1912), he resigned to establish a rubber research laboratory where he polymerized vinyl chloride in solution. He made a detour from polymer science research and investigated chemotherapy (1914-1915) but returned to the investigation of rubber in 1915. His principal contribution in USSR was the development of a commercial process for conversion of ethanol to butadiene.

When czarists became unpopular in 1921, he emigrated to the US via Latvia. As a research chemist for the US Rubber Company, he developed processes for the production of styrene from ethylbenzene and styrene-butadiene copolymers. His attempts to start pharmaceutical firms viz, Ostro Research and Pyridium Corp. were unsuccessful. In addition to his publications and patents in polymer science he authored a book on Scientific Basis for Chemotherapy in 1926.

He became an American citizen in 1930 and died in New York City on June 16, 1939.

ROY PLUNKETT

Dr. Roy Plunkett and Paul Flory attended the same universities (Manchester and Ohio State) and each was employed by E.I. DuPont de Nemeurs and Co. As outlined in his profile,, Paul left industry to join the faculty of the University of Cincinnati in 1938 and subsequently accepted positions with other universities as well as with industrial firms before accepting a chair at Stanford. Dr. Flory made many contributions to polymer science but he will be remembered most for his theoretical concepts.

In contrast, Roy spent his entire career (39 years) with DuPont and was recognized as being the first to demonstrate the production of polyfluorocarbons. Dr. Plunkett in cooperation with Jack Rebok discovered solid polytetrafluorethylene (Teflon) in a cylinder which had been filled with gaseous tetrafluorethylene. This discovery, not only provided an unusually heat and corrosion resistant product but also paved the way for the development of many other polyfluorocarbons.

Roy is a member of the American Chemical Society, the American Institute of Chemists, American Institute of Chemical Engineers, the American Association for the Advancement of Science and Sigmi Xi and has been awarded honorary doctoral degrees by Ohio State University, Washington College and Manchester College. He is the recipient of the John Scott Medal (1951), the National Association of Manufacturers Modern Pioneer Medal (1965), and the Chemist Pioneer award of the American Institute of Chemists (1964). He was inducted into the Plastics Hall of Fame (1973), named a Pioneer in Polymer Science in Polymer News (1985) and inducted into the Inventors Hall of Fame (1986).

Roy is the father of two sons by his first marriage. He married Lois M. Koch in 1965. The Plunketts, who are ardent golfers and fishermen, reside on the golf course at Padre Island Country Club near Corpus Christie, TX.

EUGENE GEORGE ROCHOW

Eugene George Rochow, who is recognized as one of the pioneers in silicone science, was born in Newark, NJ on October 4, 1909. He received the B. Chem. and Ph.D. degrees from Cornell University in 1931 and 1935.

Prior to joining the faculty at Harvard University in 1948, he spent 11 years synthesizing and commercializing silicones at GE. He was awarded eight patents by the U.S. Patent Office and is the author of 170 reports in scientific journals.   he has also published 10 books including The Chemistry of Silicones, The Chemistry of Organometallic Compounds and The Metalloids. Two of his books were coauthored by his brother Theodore of American Cyanamid.

Dr. Rochow is a fellow of the American Institute of Chemists, a member of the American Chemical Society, American Association for the Advancement for Science, and Society de Chemie Industrielle. He was awarded an honorary A.M. degree by Harvard and an honorary D.Sc. degree by Braunschweig University. He was named a Pioneer in Polymer Science by Polymer News (1988), and is the recipient of the following awards: Baekeland Medal (1941), Myer award (1951), Matiello Lecture award (1958), Perkin Medal (1961), Honor Scroll of the American Institute of Chemists (1964), Kipping award (1965), the Chemist Pioneer award from the American Institute of Chemists, Chemical Manufacturers Association Catalyst award (1971), GE Inventors award (1971) and the James Flack Norris Education award (1973).

Eugene married Priscilla Ferguson in .
Rochows were parents of Stephen, Jennifer a
Jr.   After Priscilla's death in 1950, Dr. Roch
Helen Louis Smith.   Since retirement from I
Professor Emeritus in 1970, he and Helen have   ...u at
Captiva, FL.

PAUL SCHLACK

As stated in his profile, Wallace C. Carothers is
recognized as the inventor of the first synthetic polyamide
fiber viz nylon 66.   However, S. Gabriel synthesized a
polyamide via nylon 6 in 1899. In the 1930's Paul Schlack
repeated this synthesis and produced commercial nylon 6
by the polymerization of caprolactam.

In contrast to nylon 66, which was developed by a
team of experts at DuPont, nylon 6 (Perlon L) was
developed as a side line investigation by Paul Schlack and
a part time assistant.   Like nylon 66, nylon 6 is produced
throughout the world and is used both as a fiber and an
engineering polymer.

Paul Schlack was the son of Johanna Herzog and
Theodore Schlack. He was born on December 22, 1897 in
Stuttgart where he received his education.   He was
coauthor of two textbooks on polyamides and was awarded
the A. von Bayer Medal and the Gold Medal of the
American Association of Textile Chemists and Colorists in
1953. Unlike Dr. Carothers who died before nylon 66 was
produced commercially, Paul Schlack lived to see nylon 6
become one of the world's leading fibers and engineering
polymers.

CHARLES SADRON

Many polymer scientists have used homogenous analysis for the determination of molecular size via limiting viscosity numbers and translational and rotary diffusion constants but few recognize that these techniques and investigations of birefringence of polymer solutions were pioneered by Charles Sadron.

Charles Sadon was born in Chateauroux, France in 1902. After graduating from the University of Portiers, he taught high school for a few years before receiving his Ph.D. degree from the University of Strasbourg in 1932. After postdoctoral studies at Cal Tech (1933-1934) under the direction of Dr. von Karman, he accepted an appointment on the faculty at Strasbourg.

He added polymers to liquids to increase their birefringence and showed that the increase in birefringence was related to the orientation of flexible anisotropic macromolecules. His attempts to extend his investigations to DNA and cellulose nitrate were interrupted by World War II.

However, he was able to continue this work after the end of the war in the Centre de Recherches sur les Macromolecules (CRM) which he directed in Strasbourg during the following two decades (1947-1967). Many advances in osmometry, translational diffusion, streaming birefringence and the Kerr effect were made under his direction at CRM.

In 1963, he was appointed professor of biophysics at the Museum National d'Histoire Naturelle in Paris and in 1966, he was appointed chairman of the Centre de Biophysique Moleculaire at Orleans where he remained until his retirement in 1975.

GEORGE STAFFORD WHITBY

One of the first books on plantation rubber was authored in 1920 by George Stafford Whitby who, after receiving the B.S. degree from the Royal College of Science in London in 1907 served as a chemist for Societe Financiere des Caoutchoucs in Malaya. He then entered McGill University where he obtained the M.S. and Ph.D. degrees in 1918 and 1920. He joined the faculty at McGill and was appointed a full professor in 1923. He left McGill in 1929 to accept the directorship of the chemical division of the National Research Council of Canada Laboratories (NRC).

George was born in Hull, England on May 26, 1887. His interest in rubber was catalyzed when he served as a demonstrator for Sir William Tilden after graduating from the Royal College of Science.

Dr. Whitby left NRC in 1939 to accept a comparable post in England but resigned this position and immigrated to the U.S. to join the faculty of the University of Akron as a professor in 1942. He retired in 1954 and became an American citizen in 1946.

He published over 100 reports on polymer chemistry and according to Dr. Maurice Morton, Whitby was the most honored rubber chemist of all time. Rubber World called him "a man for all seasons."

He was the recipient of the Colwyn Gold Medal (1929), the Goodyear Medal (1955) and was elected to the Rubber Science Hall of Fame. He died at Delray Beach, FL on January 10, 1972.

CONCLUSION

In spite of the accomplishments of the Polymer Pioneers cited in this and previous chapters, much remains to be discovered. Fortunately, the Center for History of Chemistry (CHOC) at the University of Pennsylvania is recording many of these new developments. The American Chemical Society is also establishing a museum showing developments in chemistry at the Smithsonian Institute. Hence, visitors to CHOC or the Smithsonian will be able to observe that the establishment of polymer science as a major discipline was the result of continued efforts by many pioneers in this important scientific field.

NAME INDEX

Aachen, Technical University of 68, 219, 229
Abderhalden, E. 17, 26
Academie des Sciences, Institut de France 109
Academy of Sciences, French 53
Academy of Medicine, French 54
Academy of Sciences, Prussian 70
Adams, R. 127, 131, 143, 174, 175, 182
Advancement of National Industry, Society of 53
Agricultural Society of France, Central 52–55 *passim*
Aix-la-Chapelle, University of 72
Akademischer Pharmazeutenverein Zurich 106
Akron, University of 249
Albert, K. 88, 90
Aleman, J. 149
Alfrey, T. 37, 148, 165
Allen, G. 10
Allen, I. 148
Allied Chemical Company 231
Alpha Chi Sigma 185, 241
Alyea, N.H. 78
American Academy of Arts and Sciences 150, 170, 174
American Academy of Achievement 185
American Association for the Advancement of Science 170, 185, 228, 229, 245, 246
American Association of Textile Chemists and Colorists 247
American Ceramic Society 199
American Chemical Society 10, 11, 28, 47, 55–60, 72, 78, 81, 86, 121, 125, 131, 141, 150, 169, 170, 171, 173, 174, 177, 185, 186, 187, 195, 197, 198, 199, 209, 223, 224, 228, 229, 230, 231, 233, 240, 241–246 *passim*, 250
American Crystallographic Association 198

American Cyanamide Corporation 88, 246
American Institute of Chemical Engineers 121, 185, 245
American Institute of Chemists 150, 170, 174, 185, 186, 187, 223, 224, 228, 229, 231, 234, 240, 241, 245, 246
American Institute of Physics 229
American Philosophical Society 170
American Physical Society 125, 170, 198
American Society of Metals 229
American Society for Mechanical Engineers 229
American Society of Biological Scientists 241
American ... see further under U.S.A.
Anderson, R.A. 55
Anghiera, P.M. d' 19
Ant-Wuorinen 37
Arcadia Institute 198
Archer, F.S. 6
Aristophanes 2
Arizona, University of 127, 173, 175
Armed Forces Chemical Association 185, 186
Armour Packing Company (Buenos Aires) 112
Arnowitz, R. 227
Arthur, J.C. Jr. 61
Arvin, J.A. 135
Asbury 36, 37
Asimov, I. 55
Association of Research Directors 185
Atlas Mineral Products Company 115
Atlas, S.M. 60
Atomic Energy Commission 133
Ault, W.C. 190
Auwers, K. von 202
Aylesworth, J.W. 86

Badger, R.M. 162
Badische Anilin und Soda-Fabrik (BASF) 68, 69, 72